Springer Theses

Recognizing Outstanding Ph.D. Research

Aims and Scope

The series "Springer Theses" brings together a selection of the very best Ph.D. theses from around the world and across the physical sciences. Nominated and endorsed by two recognized specialists, each published volume has been selected for its scientific excellence and the high impact of its contents for the pertinent field of research. For greater accessibility to non-specialists, the published versions include an extended introduction, as well as a foreword by the student's supervisor explaining the special relevance of the work for the field. As a whole, the series will provide a valuable resource both for newcomers to the research fields described, and for other scientists seeking detailed background information on special questions. Finally, it provides an accredited documentation of the valuable contributions made by today's younger generation of scientists.

Theses are accepted into the series by invited nomination only and must fulfill all of the following criteria

- They must be written in good English.
- The topic should fall within the confines of Chemistry, Physics, Earth Sciences, Engineering and related interdisciplinary fields such as Materials, Nanoscience, Chemical Engineering, Complex Systems and Biophysics.
- The work reported in the thesis must represent a significant scientific advance.
- If the thesis includes previously published material, permission to reproduce this must be gained from the respective copyright holder.
- They must have been examined and passed during the 12 months prior to nomination.
- Each thesis should include a foreword by the supervisor outlining the significance of its content.
- The theses should have a clearly defined structure including an introduction accessible to scientists not expert in that particular field.

More information about this series at http://www.springer.com/series/8790

Yu-Chuan Lin

Properties of Synthetic Two-Dimensional Materials and Heterostructures

Doctoral Thesis accepted by Pennsylvania State University, State College, PA, USA

Yu-Chuan Lin
Center for Nanophase Materials Sciences
Oak Ridge National Laboratory
Oak Ridge, TN, USA

ISSN 2190-5053 ISSN 2190-5061 (electronic)
Springer Theses
ISBN 978-3-030-13105-0 ISBN 978-3-030-00332-6 (eBook)
https://doi.org/10.1007/978-3-030-00332-6

This Springer imprint is published by the registered company Springer Nature Switzerland AG
The registered company address is: Gewerbestrasse 11, 6330 Cham, Switzerland

Supervisor's Foreword

Two-dimensional (2D) materials are arguably one of the most popular research fields in solid-state materials science over the past two decades. This material system is a true "2D" building block, thinner than 1 nm, with wafer-scale lateral dimensions that could provide the foundation for next-generation materials engineering. Their versatility for material functionalities and heterogeneities is interesting and inspires many ideas.

Furthermore, their ability to couple individual properties to generate new and novel properties has led to a wide variety of scientific breakthroughs. Most 2D materials are known as van der Waals (vdW) materials, a class of materials whose structures are highly anisotropic and whose surfaces are terminated with vdW bonds. They can be placed on any foreign surface without significant changes in their intrinsic materials properties; they can be stacked to form high-quality interfaces with other 2D materials whose lattice constant is not necessarily matched. This is where the focus of Yu-Chuan Lin's thesis research begins: with the integration of disparate 2D materials to explore novel properties that arise from their combination. Heterostructures like the ones described in Yu-Chuan's thesis have enabled hundreds of groundbreaking results in physical sciences since their first reports in 2011. In order to make vdW heterostructures technologically relevant, we must move on from mechanical exfoliation and transfer to a practical level where we can create them in an atom-by-atom, bottom-up approach.

Yu-Chuan's doctoral research at the Pennsylvania State University focuses on the growth, integration, and properties of vdW heterostructures, with an emphasis on transition metal dichalcogenides and graphene. Yu-Chuan has incorporated materials synthesis techniques, materials chemistry, and a variety of characterization techniques into his research in order to build a comprehensive study on synthetic vdW heterostructures. His graduate research on vertical vdW heterostructures out of various atomic layers has led to "firsts" in the field, including the first directly grown vdW heterostructure with epitaxial graphene, the first demonstration of novel properties in advanced heterostructures not seen before in manually stacked materials, and the first to conclusively show the importance of defects in the vertical transport of these structure. He demonstrated novel multilayer heterostructures and elucidated electronic transport and optical coupling across multiple 2D interfaces. In addition to his vdW heterostructures work, the independent investigation in his thesis also focused on epitaxial WSe_2, grown by MOCVD processes. He sufficiently elaborated the reasonings behind everything we see in MOCVD, including growth mechanisms, surface chemistry, and device performance. The implication of his thesis has advanced our understanding on the synthesis sciences and properties of nanomaterials, as the articles embedded in the content of this thesis has been cited more than 400 times. Needless to say, this thesis provides sustainable knowledge and information, as we are brainstorming to grow better 2D materials, better 2D interfaces in the future.

Materials Science and Engineering
The Pennsylvania State University
University Park, PA, USA   Joshua A. Robinson, Ph.D.

Preface

Graphene and other two-dimensional semiconductors have established a completely new research field, "2D materials" that covers all of subjects related to fundamentals and applied sciences, engineering, biology, medicine, and so forth. Their layered crystal structures and anisotropic properties are utilized to create new properties. From an electrical perspective, they are promising candidates for the low-power and flexible electronics because of their ultrathin nature, excellent electrical properties, and excellent mechanical flexibility.

While they are considered as dreamful materials by many of us, many researchers may encounter a few difficulties when studying them: The size of 2D materials prepared by mechanical exfoliation is limited. Besides the size limitation, tapes used during exfoliation usually leave polymer contamination on the surface of 2D materials. Therefore, some of researchers in this field has moved on to grow 2D materials directly on substrates, instead. Ultimately, we are hoping to synthesize 2D materials as large as we want and also be able to control properties and functionality at where exactly we want to for a variety of applications and practical use. However, several questions need to be answered first, such as the following: How to make them scalable and large area? How to control or reduce defect density of 2D materials during growth? How to integrate them with dissimilar but preferable substrates? This thesis was meant to answer some of these questions, hoping it would move the frontier of the synthesis sciences of 2D materials forwards.

This book covers two types of materials integration in the context of 2D transition metal dichalcogenides and graphene. They are (1) vertical integration of 2D layers for van der Waals (vdW) heterostructures and (2) scalable, lateral growth of WSe_2 on insulating substrates. The first two chapters cover fundamental knowledge and a brief overview on 2D transition metal dichalcogenides (TMDC) and graphene, vdW heterostructures, thin-film techniques and examples. Chapter 3 has two sections that cover the properties of synthetic WSe_2: The first is about the first demonstration of the metallic-organic chemical vapor deposition process for WSe_2, and the second covers a more sustainable process for WSe_2 on insulating substrates and also a completed study on the properties of WSe_2. Chapter 4 discusses the synthesis of MoS_2 on graphene and how morphology and quality of graphene template impact the nucleation and growth of MoS_2 and other TMDC. Chapters 5 and 6 discuss epi-taxial relationship between WSe_2 and graphene, vertical electronic transport through their heterointerface, and modulation of the carrier concentration of graphene for electrical contact. In Chap. 7, resonant tunnel diodes made of TMDC bilayer (MoS_2/WSe_2 and WSe_2/$MoSe_2$) is thoroughly discussed, including materials preparation, properties, and its electronic transport.

Oak Ridge, TN, USA   Yu-Chuan Lin

Acknowledgments

I gained valuable research experience and also obtained professional skills during my time in Department of Materials Science and Engineering at The Pennsylvania State University. After five years of hard work and some sleepless nights, I acheived one of my career objectives here: obtaining a Ph.D. degree. However, I would not have made it if there were no a good mentor and a group of wonderful and important friends who came alone at my graduate school. I would like to acknowledge Professor Joshua Robinson for offering me opportunities to exploit novel layered materials and their optoelectronic and providing me the necessary support and guidance for the success of it. He is a great mentor and always a wonderful academic father with good nature and enormous patience to me. I would like to recognize both Dr. Amy Robinson and Professor Joshua Robinson for offering me teaching assistant opportunities so I can interact with undergraduate students at Penn State, providing them short courses and laboratory instruction.

I would also like to recognize Professor Lain-Jong Li who first introduced me to materials science research when I was pursuing a master's degree in Department of Physics at National Taiwan University and also thank Professor Joan Redwing and Dr. Sarah Eichfeld for introducing me to metal-organic chemical vapor deposition for 2D semiconductors. I am thankful to my past and current colleagues at graduate school for assistance in research and insightful discussion. In particular, I would like to express my appreciation to Dr. Ganesh Bhinamapati, Brian Bersch, Kehao Zhang, Shruti Subramanian, Natalie Briggs, Jennifer DiStefano, Maxwell Wetherington, Chia Hui Lee, Lorrain Hossaine, Donna Deng, and Dr. Bhakti Jariwala for their instrumental help and collaboration within the group. I also thank my supportive collaborators outside Penn State, they are Professor Robert Wallace, Professor Susan Fullerton-Shirey, Professor Randall Feenstra, Professor Kyeongjae Cho, and their students and postdocs for unselfish collaboration and input on our collaborations in the Center for Low Energy Systems Technology (LEAST). I am also grateful to the LEAST program for its funding support for my graduate research and stipend.

There are also important friends outside my research I would like to thank to for their friendship, including Jeremy Schreiber, Ece Alat, Alperen Ayhan, and Fredrick

Lia. Especially, I would like to deliver my sincere gratitude to Birgitt Boschitsch for her great support and friendship throughout my school years. Finally, I would like to thank to my parents, Chih-An and Mei-Ying; my sister, Mei-Xuan; and all my friends in my hometown whose endless love and support, and endurance of my long-term absence enable me to pursuing my dreams on the other side of the Pacific Ocean. In particular, I would like to recognize my caring parents for introducing me the motivation, drive, and diligence that came along with me so far.

Contents

Parts of this thesis have been published in the following articles:

1. Lin Y-C, Ghosh RK, Addou R, Lu N, Eichfeld SM, Zhu H, Li M-Y, Peng X, Kim MJ, Li L-J, Wallace RM, Datta S, Robinson JA (2015) "Atomically thin resonant tunnel diodes built from synthetic van der Waals heterostructures," *Nat. Commun.* 6:7311. https://doi.org/10.1038/ncomms8311
2. Lin Y-C, Chang C-YS, Ghosh RK, Li J, Zhu H, Addou R, Diaconescu B, Ohta T, Peng X, Lu N, Kim MJ, Robinson JT, Wallace RM, Mayer TS, Datta S, Li L-J, Robinson JA (2014) "Atomically thin heterostructures based on single-layer tungsten diselenide and graphene," *Nano Lett.* 14:6936–6941. https://doi.org/10.1021/nl503144a
3. Lin Y-C, Lu N, Perea-Lopez N, Li J, Lin Z, Peng X, Lee CH, Sun C, Calderin L, Browning PN, Bresnehan MS, Kim MJ, Mayer TS, Terrones M, Robinson JA (2014) "Direct synthesis of van der Waals solids," *ACS Nano* 8:3715–3723. https://doi.org/10.1021/nn5003858
4. Lin Y-C, Jariwala B, Bersch BM, Xu K, Nie Y, Wang B, Eichfeld SM, Zhang X, Choudhury TH, Pan Y, Addou R, Smyth CM, Li J, Zhang K, Haque MA, Fölsch S, Feenstra RM, Wallace RM, Cho K, Fullerton-Shirey SK, Redwing JM, Robinson JA (2018) "Realizing large-scale, electronic-grade two-dimensional semiconductors," *ACS Nano* 12:965–975. https://doi.org/10.1021/acsnano.7b07059
5. Eichfeld SM, Hossain L, Lin Y-C, Piasecki AF, Kupp B, Birdwell AG, Burke RA, Lu N, Peng X, Li J, Azcatl A, McDonnell S, Wallace RM, Kim MJ, Mayer TS, Redwing JM, Robinson JA (2015) "Highly scalable, atomically thin WSe$_2$ grown via metal-organic chemical vapor deposition," *ACS Nano* 9:2080–2087. https://doi.org/10.1021/nn5073286
6. Lin Y-C, Li J, de la Barrera SC, Eichfeld SM, Nie Y, Addou R, Mende PC, Wallace RM, Cho K, Feenstra RM, Robinson JA (2016) "Tuning electronic transport in epitaxial graphene-based van der Waals heterostructures," *Nanoscale* 8:8947–8954. https://doi.org/10.1039/C6NR01902A

Chapter 1
Two-Dimensional Materials

1.1 Introduction

Size effect can dictate the properties of the materials. At the nanoscale, changing the number of atoms and molecules forming the materials leads to qualitative changes in physical and chemical properties because the length of interaction from one atom (molecule) to another is approaching to the size of the entire materials. One well-known example of size-dependent phenomena is the quantum confinement effect in ultrasmall semiconducting materials [1]. The term "nanomaterial" is used to describe the materials that have at least one of their dimension in the nanometer scale. Prior to the 1980s, nanoscale materials and technology was only conceptual (i.e., the lecture "There is plenty of room at the bottom" by Richard Feynman in 1959 and the term "nanotechnology" proposed by Norio Taniguchi in 1974) [2, 3] because manipulating atoms and molecules of the materials precisely and achieving high-resolution images in the small scale were difficult at the time. Besides experimental challenges, it was commonly acceptable that a material in such scale may not be stable in room temperature due to large atomic displacement caused by thermal fluctuation. Even Feynman himself also claimed in his lecture that glass and plastic are better candidates than metal and crystals for machines and electronics in the small scale because the later ones will separate into domains to make their lattice structure stronger [2].

Thanks to the rapid development in the field of surface probe technique, including scanning tunneling microscopy invented by Gerd Binnig and Heinrich Rohrer in the early 1980s [4], the public and science community was able to look into colloidal and interface sciences more effectively. For the nanomaterials, this breakthrough in surface science began at the exploration of fullerene (C60) in the 1980s [5], carbon nanotubes in the early 1990s [6], and continued all the way to semiconducting quantum dots in the late 1990s [7].

© Springer Nature Switzerland AG 2018
Y. -C. Lin, *Properties of Synthetic Two-Dimensional Materials and Heterostructures*, Springer Theses,
https://doi.org/10.1007/978-3-030-00332-6_1

At the time they were explored, each class of these nanomaterials exhibits unprecedented properties arising from their dimensionality. If one dimension is restricted, a layered shape or 2D material can be made; if two dimensions are limited in size, a wire or 1D material can be found; if all dimensions are in the range of a few nanometers, 0D material is then produced. The most representative case can be seen on the sp^2 carbon materials, where graphite (3D), fullerenes (0D), nanotubes (1D), and graphene (2D) were typically presented in the chronological order of their earliest findings. Driven by the novelty that nanomaterials provide, and also significantly by ample curiosity, scientists, particularly solid-state physicists, had been trying to thin down graphite aggressively by many means, including intercalation or rubbing graphite on a substrate [8, 9]. However, the properties of monolayer hexagonal carbon atoms were not fully explored until it was successfully isolated and thoroughly characterized by Geim and co-workers at the University of Manchester in 2004 [10]. Beyond Feynman and other scientists' understanding, not only graphene can be stable on a substrate at room temperature and have high crystal quality but also exhibits unprecedented properties that are very different to its counterparts in another dimensionality.

This single event in 2004 was indeed the birth of the field of two-dimensional materials, and its preparation method also triggers the exploitation of other non-graphene 2D layered materials, like hexagonal boron nitride (hBN), transition metal dichalcogenides (TMDC), and 2D black phosphorus (phosphorene) [11, 12]. 2D materials exhibit numerous exceptional properties. First, quantum confinement effect in the direction perpendicular to the basal plane leads to unprecedented electronic and optical properties that are absent in their parental crystals [12, 13]. Second, unlike traditional 3D materials such as gallium arsenide (GaAs) and silicon (Si), their surfaces are free of dangling bonds and their structure is mechanically robust and henceforth makes it easy to integrate 2D materials with functional structures such as cavities and photonic crystals. In addition, its van der Waals interaction enables 2D materials to construct a vertical heterostructure without suffering the lattice mismatch issues when using layers with different lattice constants. Third, the light-matter interaction in many 2D semiconductors is strong, despite their atomic thickness (i.e., 1 L MoS_2 absorbs 10% of vertically incident light at its excitonic resonance) [14]. The energy bandgap of these 2D layers constituted a continuous energy spectrum in the ranges from infrared to visible wavelength, as shown in Fig. 1.1. In other words, 2D materials not only extend the frontier of fundamental science but also serve as components in optoelectronics and photovoltaic devices used in our daily life.

The rapid growth of the field of 2D materials can be reflected by the increasing number of publication within the last 10 years (Fig. 1.2) [15]. Especially, after graphene gained international attention at the Nobel Prize in 2010, the number of research papers related to graphene increased by tens of thousands annually. Similarly, papers that reported other 2D layers including TMDC, monochalcogenides (i.e., InSe, GaSe, etc.), monoelemental 2D semiconductors (i.e., silicene, phosphorene, germanene), and MXenes also have been steadily increasing since 2011. Needless to say, 2D materials have become indispensable in academia and will soon be utilized in every aspect of our daily life in the near future.

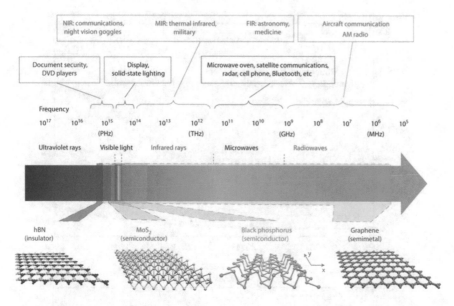

Fig. 1.1 The electromagnetic wave spectrum and the bandgap ranges of various types of 2D materials. NIR, MIR, and FIR indicate near-, mid-, and far-infrared, respectively. The atomic structures of hBN, MoS_2, black phosphorus, and graphene are shown in the bottom of the panel, left to right. The crystalline directions (x and y) of anisotropic black phosphorus are indicated [12]

1.2   Classification and Thermal Stability

Two-dimensional materials exhibit fundamental properties that are absent in their bulk counterparts. After the successful isolation of graphene, which is one-carbon-thick layer, the research that focused on 2D materials has been growing rapidly with an eye toward applications in semiconductors, energy harvesting, electrodes, and membranes for water purification. In order to fully explore the total number of 2D materials, 65,000 inorganic crystal compounds with crystallographic and thermodynamic data in online the Materials Project (MP) database had been examined by Hennig and co-workers using topology-scaling algorithm (TSA) to verify layer compounds [16]. TSA is able to simultaneously identify bonded networks of any dimension and classify structures in the MP database systematically. One task of TSA is to identify structural patterns that are separated from each other by distances larger than the bond length of atoms within the pattern [16]. Their theoretical efforts identified 826 stable or semi-stable layered materials (LM). According to their stoichiometric ratios, 826 2D materials can be grouped into several categories. Among those, more than 50% of LM are presented by AB_2, ABC, AB, AB_3, and ABC_2, in decreasing frequency inversely proportional to their complexity (Fig. 1.3a). The percentages of unary, binary, ternary, and more complexed types of stable layered compounds are compared with the percentages of all stable compounds. It shows that binary, ternary, and quaternary compounds comprise ~ 98% of the stable

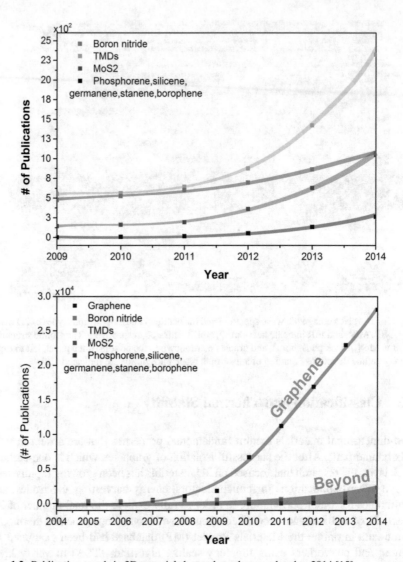

Fig. 1.2 Publication trends in 2D materials beyond graphene updated to 2014 [15]

layered compounds. In addition, the percentages of unary and ternary compounds among layered materials are close to their percentages among all materials in the MP database (Fig. 1.3b, c). The thermodynamic stability of layered materials determines if their 2D counterpart can be isolated. They are always metastable and not true thermodynamic ground states, as the total energy of the system is always lower when two layers are brought together. Despite the oppositions from theoretical perspective, 2D materials have been proved to be kinetically stable by themselves. The thermodynamic stability of a 2D material can be described by the difference in the energy of a 2D material and the lowest value for its bulk part [17]:

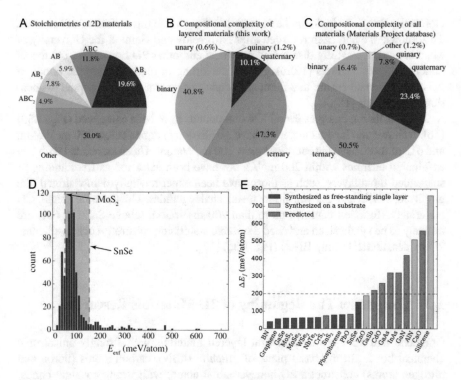

Fig. 1.3 The classification and stability of 2D materials. (**a**) Distribution of stoichiometry of 826 layered compounds whose presence has been theoretically confirmed. The top 5 most common stoichiometry are ABC, AB2, AB, AB3, and ABC2 that represent half of all compounds. (**b**) Distribution of comparison of the compositional complexity among the stable layered materials identified by this work. (**c**) Distribution of all materials in the MP database. The percentage of binary compounds (one cation and one anion) among layered materials is significantly higher than among all materials. It indicates that binary compounds are relatively conducive to creating inter-layer interactions. (**d**) Histogram of calculated exfoliation energies for 826 layered materials is compared to the range of calculated exfoliation energies for synthesized 2DM. Most of compounds have the energies below 100 meV/atom and could be easily exfoliated. (**e**) All 2D materials that have been synthesized as freestanding films have formation energies below 200 meV/atom, illustrated by the horizontal dashed line [16, 17]

$$\Delta E_f = \frac{E_{2D}}{N_{2D}} - \frac{E_{3D}}{N_{3D}},$$

where E_{2D} and E_{3D} are the energies of the single-layer and bulk (or mixture of bulk) materials, respectively, and N_{2D} and N_{3D} denote the numbers of atoms in the respective unit cells. In general, the lower the exfoliation energies of 2D materials are, the higher their thermodynamic stability would be. Previously, the materials in 2D form were presumably impossible because the theory stated that the thermal fluctuations in low-dimensional crystal lattices would make displacement of atoms exceed their interatomic distances at any temperatures [18]. Owing to this reason, atomically thin films had only been considered as an integral surface of 3D crystals. This knowledge had been revisited after Novoselov and Geim at the University of Manchester successfully discovered graphene and other 2D crystals on the top of noncrystalline substrates [10]. Arguably, the strong in-plane interatomic bonds of 2D crystals would ensure that thermal fluctuations cannot generate crystal dislocations and defects [18].

The exfoliation energies for all 826 compounds have been calculated (Fig. 1.3d) [16]. It shows 680 layered materials have exfoliation energies below 150 meV/atom and 612 of those have the energies below 100 meV/atom. Those associated with low exfoliation energies within 200 meV/atom have been extracted as freestanding or suspended monolayers, such as those have been experimentally demonstrated like graphene, hBN, and 2D transition metal chalcogenides. On the other hand, 2D materials exfoliation energies more than 200 meV/atom (above SnSe in (D)) are unlikely to be synthesized and need a suitable stabilizing substrate, such as silicene, 2D oxides, and 2D group III−N (Fig. 1.3e) [17].

1.3   Graphene: The Beginning of 2D Materials Research

Graphene and hBN are isostructural layered materials with strongly anisotropic chemical bonds. In the basal plane of graphite (hBN), carbon atoms (boron and nitrogen atoms) construct a 2D honeycomb structure with strong covalent bonds, while the basal planes interact weakly with each other via van der Waals (vdW) bonds (Fig. 1.4). Therefore, many of their physical properties, such as energy band structure, electrical conductivity, thermal conductivity, Debye temperature, phonon type, and magnetism type, are highly anisotropic. Their basal plane has surface energy much lower than the other surfaces have due to the absence of dangling bond and, hence, makes integration of basal planes with various solid surfaces possible. From a mechanical aspect, the abovementioned 2D bonds exhibit a high flexibility for bending their basal plane. The restoring force for mild bending of the basal plane

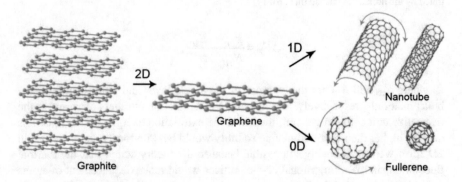

Fig. 1.4 Graphite (3D) and its one basal plane (2D). Graphene can be further rolled into 1D and 0D structures [18]

is smaller than that of 3D crystals because the polarized transverse acoustic mode normal to the basal plane has parabolic dispersion the regime of longer wavelength, instead of the linear one that is common in 3D crystals [18]. This mechanical robustness consequently enables the 0D and 1D nanostructures of graphite and hBN such as C60 and multiwall nanotubes made by rolling their basal planes. The epitaxial films of ultra-thin graphite and hBN had been studying for decades before their 2D allotrope, owing to their interesting characters [19]. For example, a thin film of graphite was grown on transition metal substrates (i.e., Ni, Pt, Ir, Pd) and carbide substrates (i.e., TiC, TaC, HfC) via chemical vapor deposition techniques and subsequently probed by low-energy electron diffraction, Auger electron spectroscopy, and high-resolution electron energy-loss spectroscopy for understanding phonon dispersion, band structures, epitaxial relationship, etc. [19].

Geim and Novoselov used the "Scotch tape" for graphite exfoliation in order to create ultra-thin graphite layers on an insulating substrate. Tapes have been used to clean residue off TEM grids and happen to provide just enough force to decouple vdW interaction between graphene layers in graphite (Fig. 1.5a) [18]. An atomically thin layer was successfully extracted by the exfoliation technique and its size is typically ranging from sub-μm to 100 μm (Fig. 1.5b) [18]. In graphene, each carbon atom provides three electrons that bound with the nearest-neighbor electrons, thus creating a covalent bond (sp^2). For each atom, a fourth electron (π) is delocalized on the whole graphene, which enables the conduction of current. If the energy of the electron is represented in function of their momentum, the bands are in a parabola shape. The energy bands form two circular cones, connected one with the other at their extrema. They are called Dirac cones (Fig. 1.5c, d) [20]. Graphene presents an uncommon behavior because it does not have a gap, unlike the insulators, but also no partially filled band, unlike metals. Layered materials enabled the realization of pure 2D systems and present peculiar phenomena. While a traditional 2D electron gas (2DEG) is confined to the interface of two tandem epitaxial III–V semiconductors [22], graphene has been regarded as "real" 2DEG system. There are numerous interesting phenomena raising from its Dirac cone band structure, such as ambipolar transfer characteristics, a mobility of 10^6 cm^2/Vs at room temperature, and only 2% absorbance in the whole range of visible wavelength (Fig. 1.5e, f) [18, 21]. Despite the great advantages that it provides to science and engineering community, its gapless nature raises the concerns for realization of graphene-based digital applications [23].

1.4 Monolayer Transition Metal Dichalcogenides: Real 2D Semiconductors

Semiconducting 2D TMDC came into the play and made the applications and science that graphene cannot achieve possible because they have a sizeable bandgap. They have the common chemical formula MX_2 where M is for a transition metal (i.e., Mo, W, Ta) and X is for S, Se, or Te atoms. Bulk TMDC crystals are formed by

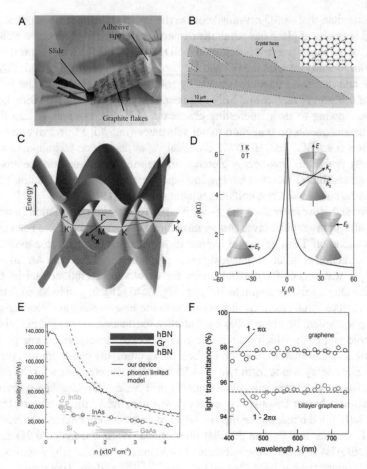

Fig. 1.5 (a) The "Scotch tape" procedure isolates graphene on a substrate, reported by Novoselov and Geim in 2004. (b) Scanning electron microscopic image of a graphene, which shows that most of its faces are zigzag and armchair edges as indicated by blue and red lines and illustrated in the inset. (c) Band structure of graphene shows conductance band touches the valence band at the K and K' points, which is so-called the Dirac point. (d) Ambipolar FET in graphene. The insets show its conical low-energy spectrum $E(k)$, indicating changes in the position of the Fermi energy E_F with changing gate voltage V_g. Positive (negative) V_g induce electrons (holes) in concentrations $n = \alpha V_g$ where the coefficient α depends on the use of dielectrics (7.2×1010 cm^{-2} V^{-1} 300 nm SiO$_2$). The rapid decrease in resistivity ρ on adding charge carriers indicates their high mobility (in this case, $\mu \approx 5000$ cm^2 V^{-1} s^{-1}). (e) Mobility versus density at room temperature (solid black curve). Dashed black curve indicates the theoretical mobility limit due to acoustic-phonon scattering. Graphene FET is in comparison with the range of nobilities reported in other semiconductors. The inset shows that both sides of graphene have been encapsulated by hBN. (f) Transmittance spectra of single and bilayer graphene show that every one layer absorbs 2.3% of incident white light as a result of graphene's electronic structure [18, 20, 21]

vertical stacking of monolayers separated by ~6.5 Å. One monolayer contains a three-layer stack of X-M-X (Fig. 1.6a) [24]. Mono- and few-layer flakes of TMDC can be easily extracted from bulk crystals in the same way that graphene was deposited on a cleaned substrate (Fig. 1.6b, c). The bandgap of bulk TMDC crystals is around 1 eV but can further increase to 1.5–2.2 eV once they are thinned down to monolayer. O. Yazyev and A. Kis in their review have shown MoS_2 is an indirect gap semiconductor with valence band maximum (VBM) located at the Γ-point and conduction band minimum (CBM) located at a low-symmetry point of the Brillouin zone [24]. Upon thinning MoS_2 layers, the shapes of valence and conduction bands undergo changes, such that the positions of both of its VBM and CBM shift to the K-point making an indirect-to-direct bandgap crossover.

The change in the band structure with layer number is due to quantum confinement and the resulting change in hybridization between p_z orbitals on S atoms and d orbitals on Mo atoms. The electronic distributions are also spatially correlated to the atomic structure. Density functional theory (DFT) calculations for MoS_2 (Fig. 1.6d, f) show that the states of conduction band at the K-point are mainly introduced by localized d orbitals on the Mo atoms. These states are located in the middle of the "S-Mo-S" sandwiches and relatively intact to interlayer coupling [24]. On the other hand, the states near the Γ-point are the combined efforts of the antibonding p_z orbitals on the S atoms and the d orbitals on Mo atoms and have a strong interlayer coupling effect. Therefore, as the layer numbers change, the direct excitonic states near the K-point are relatively unchanged, but the transition at the Γ-point shifts significantly from an indirect one to a larger, direct one. All semiconducting MX_2 compounds are expected to undergo a similar transformation with decreasing layer numbers, covering the bandgap that ranges from 1.1 to 1.9 eV.

Significant efforts to open the bandgap of graphene using graphene nanoribbons, AB-stacked bilayer graphene, and chemical doping (i.e., substituting C of graphene with B and N) had negligible success, providing the bandgap opening up to 200 meV as the best [25]. This challenge serves as a driving force in developing 2D TMDC with a finite bandgap. 2D TMDs reveal a wide range of bandgap covering all visible and infrared range with the choice of material. Most semiconducting 2D TMDC reveal direct bandgap in monolayer, whereas their bulk counterparts are indirect bandgap (exceptional cases are InSe and $ReSe_2$). For example, 2D MoS_2 (1.8 eV), $MoSe_2$ (1.5 eV), (2H)-$MoTe_2$ (1.1 eV), WS_2 (2.1 eV), and WSe_2 (1.7 eV) show direct bandgap [24]. Depending on the structures, constituent elements, and amounts of electron in d orbitals of transition metal elements, 2D TMDC layers can exhibit metallic/semiconducting behaviors, charge density wave (CDW), magnetism (ferromagnetic and antiferromagnetic), and superconductivity (Fig. 1.7) [26]. The tremendous diversity of their properties indeed enriches the knowledge of the solid-state physics and enables numerous applications.

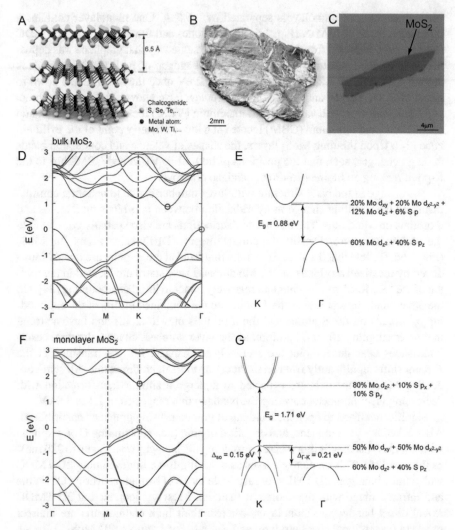

Fig. 1.6 (**a**) Schematic representation of the structure of a TMDC material with a formula MX$_2$ where metallic atoms are shown in black and chalcogens (X) in yellow. (**b**) Photograph of a bulk crystal of MoS$_2$ that can be used as a starting point for the exfoliation of single layers. (**c**) Optical image of a monolayer MoS$_2$ deposited on the surface of SiO$_2$. (**d**, **e**) Electronic band structures of bulk MoS$_2$ and monolayer MoS$_2$ calculated from first principles using density functional theory (DFT) within the generalized gradient approximation (GGA). Valence band maxima (VBM) and conduction band minima (CBM) are indicated by red and blue circles, respectively. Energies are given relative to the VBM. Schematic drawings of low-energy bands in (**f**) bulk MoS$_2$ and (**g**) monolayer MoS$_2$ showing their bandgaps (E_g) as well as the valence band spin-orbit splitting Δso and the Γ-valley band offset $\Delta\Gamma$-K for the case of monolayer MoS$_2$. The band structure parameters have been obtained at the DFT-GGA level of theory. The orbital composition of electronic states at band extrema is indicated [24]

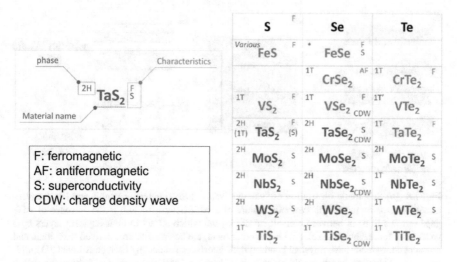

Fig. 1.7 Summary of various physical properties from a variety of monolayer TMDC [26]

1.5   2D Materials as the Building Blocks for vdW Heterostructures

A heterostructure consists of different electronic materials and has a varied energy gap. In principle, a heterostructure utilizes its energy gap variations to control electrons and holes in terms of their flow and distribution, in addition to electrical fields. Since the proposed design principle of heterostructure devices in 1957 by Kroemer [27], it has been an essential requirement for high-performance transistors, semiconducting lasers, and optical devices made out of conventional semiconductors [28, 29]. Graphene and beyond-graphene layered materials, especially atomically thin TMDCs, have created a vast field that generates more than a thousand of publications on the study of their fundamentals and material applications each year [15, 30]. These publications provide throughput fundamental understandings on every aspect of each layered materials and enable people to select specific layered materials for their needs. While new opportunities of discovering exotic phenomena in one layered material itself are running low, a new focus going beyond this field has been initiated. Various isolated monolayers of TMDCs and other vdW crystals are assembled into a sophisticated structure made into a layer-by-layer sequence that is purposely designed. These vdW heterostructures have been synthesized and investigated extensively since 2010 and already revealed new properties and exotic phenomena yet presented in their constituent layers [30, 31]. While most of ultra-thin layered crystals have been explored and demonstrated in optoelectronics, the emerging vdW heterostructure is raising a "layered renaissance" for the next-generation devices [32–35] (Fig. 1.8).

Van der Waals (vdW) heterostructures consist of a variety of 2D layered crystals that have strong in-plane covalent bonds and weak out-of-plane vdW interaction

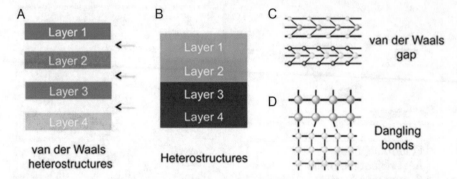

Fig. 1.8 (**a**) Schematic illustration of "van der Waals" (vdW) heterostructures and (**b**) "conventional" heterostructures. The symbolic feature of a vdW heterostructure is the presence of vdW gaps (arrows in **a**) in between constituent 2D layers, which attract to their adjacent layers by a weak vdW force, shown in (**c**). (**d**) On the other hand, the heterostructures derived from ionic and covalent compounds have physical bonds at their interfaces connecting each constituent 3D building block. Dangling bonds would be caused if there's a large lattice mismatch between grown material and growth template in the 3D cases [36, 37]

[34, 36]. The key feature that distinguishes vdW heterostructures (Fig. 1.9a, c) [38] from conventional heterostructures (Fig. 1.9b, d) [38] is the vdW gaps presenting in between constituent layers [36]. While the convenient heterostructures derived from 3D solids, such as III–V compounds, SiGe epitaxy layers, and oxides (i.e., perovskites, spinels, and dielectrics) [39] involve covalent bonds to bridge the constituent materials [29, 36, 39], vdW heterostructures bridge their constituent layers merely with weak vdW forces. Without physical bonds involved, their interfaces can tolerate a highly lattice mismatch combination (Fig. 1.9c).

1.5.1 Making vdW Heterostructures via Stacking Exfoliated 2D Layers

When reliable techniques to synthesis of high-quality vdW heterostructures are still under development, the simplest fabrication technique is to mechanically transfer one 2D crystal onto another in a step-by-step manipulation [36, 40]. The easiest report of this route is from Dean et al. [31, 41] on graphene and hBN stacks, in which a micromanipulator was used, under an optical microscopy, to precisely deposit graphene that is closely aligned to a hBN flake (Fig. 1.9a). The electrical transport measurement on the graphene integrated with hBN flakes shows a significant improvement in the field-effect mobility of graphene (Fig. 1.9b). These results indicate that hBN serves as a substrate better than SiO_2 for graphene electronics due to their closely matched lattice constants, an atomically flat surface, and lack of dangling bonds. The stacking methods can apply to layered materials that are not structurally compatible or unlikely can be grown on each other. This method had

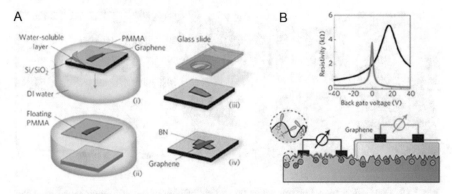

Fig. 1.9 (**a**) Schematic flow of the transfer process used to deposit exfoliated graphene on hBN flakes. (**b**) Top: The field-effect mobility of graphene-hBN devices achieved 60,000–80,000 cm^2 V^{-1} S^{-1}, a significant improvement compared with graphene-SiO$_2$ devices. Bottom: The atomically flat and dangling bond-free surface of hBN attribute to the success of the high mobility [31, 37]

inspired numerous works that built exotic vdW heterostructures to discover the new properties and applications [30, 42, 43]. There have been many new prototypes of vertical devices made from stacked exfoliated layers, while the synthesis techniques for ultra-high-quality heterostructures are under development.

1.5.2   Applications of vdW Heterostructures for Electrical and Optical Devices

There are a wide variety of devices demonstrated with vdW heterostructures utilizing a variety of 2D crystal categories including conductors, insulators, and semiconductors. The groundbreaking work by Britnell et al. [43, 44] utilized hBN flakes ranging from five to seven layers as a tunneling barrier between two sheets of graphene serving as the top and bottom electrodes in the vertical field-effect tunneling transistors in hBN-Gr-hBN-Gr-hBN (Gr: graphene) vertical heterostructures (Figure 1.10a–c). The amount of tunneling current density of the vertical devices can be tuned by controlling finite doping density and applied bias (Fig. 1.10b–g). The transistors show a tunneling I–V characteristics and orders of the on/off ratio, which address the weakness of planar graphene field-effect transistors due to lack of on/off ratio. In addition to rigid devices, flexible devices and technology utilizing 2D crystals are also emerging. The strain limit of thin-film devices made of TMDCs and other monolayers possesses a value 3–5 times greater than that made from III to V compounds, metal oxides, and crystalline silicon. Similar to the first prototype of Gr-hBN-Gr devices, Georgiou et al. [45] prepared a Gr-WS$_2$-Gr vertical heterostructure fabricated on a flexible polyethylene terephthalate films using the same transfer methods (Fig. 1.11a, b). The device exhibits

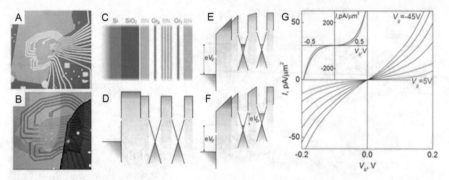

Fig. 1.10 Field-effect tunneling transistors on vertical hBN-graphene-hBN-graphene-hBN heterostructure: (**a**) optical image of the final device. (**b**) Electron microscopic image captured prior to evaporation of the Au electrodes shows two Hall bars made from graphene are shaded in green and orange. (**c**) Schematic structure of the experimental vertical devices. (**d**) The corresponding band structure without applied gate voltage; (**e**) the same band structure subjected to a finite gate voltage (V_g) and zero bias (V_b); (**f**) both of V_g and V_b are applied. (Only the tunnel barrier for electrons is considered in the illustrations.) (**g**) Tunneling characteristics for the vertical tunneling device with five to seven layers of hBN as the tunnel barrier. I–V curves for different V_g, in a 10-V step. Due to finite doping, the minimum tunneling conductivity is achieved at V_g around 3 V. The inset compares the experimental I–V curves at $V_g = 5$ V (red curve) with theory (dark curve, which takes the linear density of state in the two graphene layers into consideration and assumes no momentum conservations [37, 44])

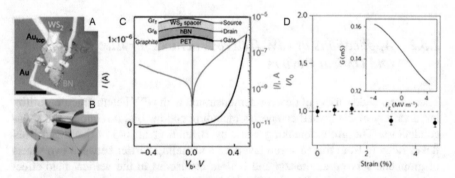

Fig. 1.11 A transistor Gr-WS$_2$-Gr built on a flexible polymer substrate: (**a**) optical image and (**b**) image of the device under bending. (**c**) I–V plot at $T = 300$ K for the bended device with $V_g = 0$. Curvature is 0.05 mm^{-1}. (**d**) Relative current variation versus applied strain. Standard variations for several consecutive measurements are shown in error bars. Inset is the gating transport of the device under strain [37, 45]

tunneling characteristics under applied characteristics (Fig. 1.11c) and maintains its electrical performance subjected to a strain up to 5% (Fig. 1.11d) [45]. The flexible structures also demonstrate a transistor effect (inset, Fig. 1.11d) and may have improved performances with the optimal thickness of dielectrics. Besides vertical tunneling devices, engineering the band structures of heterostructures can also lead to practical lightning devices using TMDs (MoS$_2$, WS$_2$, WSe$_2$) as light emitter [46,

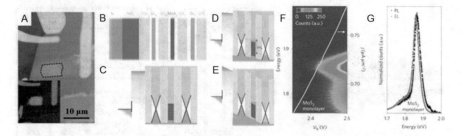

Fig. 1.12 Heterostructure devices with a single quantum well (SQW), made from hBN/Gr$_{Bottom}$/3 L hBN/1 L MoS$_2$/3 L hBN/Gr$_{Top}$/hBN, shown in (**a**), the optical image. The inset of (**a**) is electroluminescence (EL) image from the same device, under V_b = 2.5 V, T = 300 K. (**b**) The schematic structure of the same SQW device; (**c–e**) band alignment for the case of zero applied bias (**c**), intermediate applied bias (**d**), and high applied bias (**e**), for the heterostructure presented in (**b**). (**f**) EL spectra as a function of applied bias (V_b) for the SQW device made from MoS$_2$. White curve is its current density versus applied bias characteristics (j-V_b) and (**g**) comparison of the PL and EL spectra for the same devices [37, 46]

47]. The work by Withers et al. [46] fabricated single quantum well (SQW) emitters made from a stack of hBN/Gr$_{Bottom}$/3 L hBN/1 L MoS$_2$/3 L hBN/Gr$_{Top}$/hBN (Fig. 1.12a, b), in which 1 L MoS$_2$ serves as a light emitter excited by an applied bias and operates at 300 K (Fig. 1.12c–g). One of the demonstrated devices in this work achieved extrinsic quantum efficiency nearly to 10%, and the emission can be tuned over a wide range of wavelength by choosing different types and thickness of 2D semiconductors. By stacking more repetitive SQW, a multiple quantum well (MQW) vdW heterostructure with enhancement emitting intensity was realized.

Combining different TMDC monolayers can lead to a new class of van der Waals solids that exhibits new optical and electrical properties. The theoretical work by Terrones et al. [48] predicted that stacking MoS$_2$-WSe$_2$ heterostructures (Fig. 1.13a) will yield electronic properties that are entirely different from their constituent layers, such as a significantly reduced bandgap energy (Fig. 1.13b, c). This exciting theoretical work got support from many experimental works aiming to realize the theoretical results through manual stacking of different TMDC layers [50–52]. For example, a manually stacked MoS$_2$-WSe$_2$ heterostructure made by Fang et al. [50] exhibits an interlayer exciton at 1.55–1.59 eV (Fig. 1.14) [44], in addition to intralayer excitons of (1.87 eV) MoS$_2$ and (1.65 eV) WSe$_2$ monolayers [50]. More interestingly, adding electrically insulating hBN monolayers into MoS$_2$-WSe$_2$ heterostructures can modify strength of the interlayer coupling and result in decoupling of the layers, as evident by a decreased intensity of the interlayer excitons [50].

1.5.3   Interfacial Imperfection

The manual stacking process is indeed practically useful for integrating various 2D materials to create a variety of proof-of-concept van der Waals heterostructures. However, the process requires multiple steps to complete an assembly, including

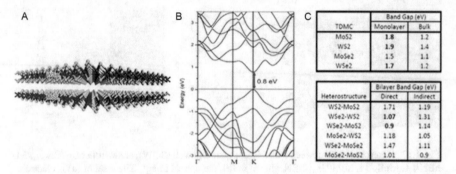

Fig. 1.13 (**a**) Simulated MoS_2-WSe_2 heterostructure yields a new direct bandgap, which is shown in (**b**). (**c**) Bandgap of monolayer and bulk TMDCs and their heterostructures [37, 49]

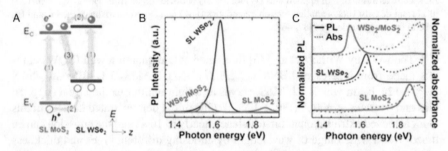

Fig. 1.14 (**a**) Band diagram of WSe_2/MoS_2 heterobilayer under photoexcitation, illustrating (1) exciton generation in 1 L WSe_2 and MoS_2, (2) relaxation of excitons at the MoS_2-WSe_2 interface where the band has been offset, and (3) radiative recombination of spatially indirect excitons. (**b**) 1 L MoS_2, WSe_2, and their heterolayer exhibit PL spectra created by radiative recombination of intralayer and interlayer excitons. (**c**) Normalized PL (solid lines) and absorbance (dashed lines) spectra of 1 L WSe_2, MoS_2, and their corresponding heterolayers, where the spectra are normalized to the height of the strongest PL/absorbance peak [37, 49]

isolating a 2D layer in micro-size, transferring it onto polymer-supporting films, stacking 2D crystals repeatedly, repeating standard clean room procedure in terms of cleaning, dissolving, resist spinning, and so on, and a precision-demanding alignment under a microscope with a micromanipulator [30, 36]. These multiple steps, carried out in ambient, unavoidably introduce contaminations at the interfaces of constituent layers (Fig. 1.15a) [45], which could be due to the presence of adsorbents and the usage of polymer films in the transfer process. Although a clean and sharp interface in these heterostructures is still obtained by confining trapped residues into "bubbles" with van der Waals forces that bond adjacent constituent layers [30, 53], a sophisticated process is still needed in order to fabricate a useful device. The visual example of the bubbles is shown in Fig. 1.15b, where the top-gate contact was deposited in a shape that the "bubbles" would be avoided [54]. In order to make van der Waals heterostructures practically useful for digital industries, an alternative for the synthetic vdW heterostructures with clean interfaces needs to come out.

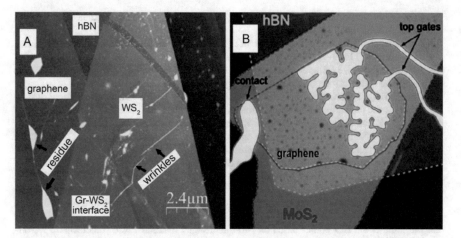

Fig. 1.15 (a) "Bubbles" and wrinkles in manually stacked van der Waals heterostructures formed due to segregated residues at their interfaces after the transfer process. Each stacked layer and their overlap is highlighted, and none of them are completely free of imperfection. (b) While fabricating devices on the heterostructures, the top contacts are shaped irregularly to avoid the polymer residue (black dots in figure) that is common in this technique [37, 54]

References

1. Takagahara, T., Takeda, K.: Theory of the quantum confinement effect on excitons in quantum dots of indirect-gap materials. Phys. Rev. B. **46**, 15578–15581 (1992)
2. Feynman, R.P.: There's plenty of room at the bottom [data storage]. J. Microelectromech. Syst. **1**, 60–66 (1992)
3. Taniguchi, N.: Current status in, and future trends of, ultraprecision machining and ultrafine materials processing. CIRP Ann. Manuf. Technol. **32**, 573–582 (1983)
4. Binnig, G., Rohrer, H.: Scanning tunneling microscopy–From birth to adolescence. Rev. Mod. Phys. **59**, 615–625 (1987)
5. Kroto, H.W., Heath, J.R., O'Brien, S.C., Curl, R.F., Smalley, R.E.: C60: Buckminsterfullerene. Nature. **318**, 162–163 (1985)
6. Iijima, S., Ichihashi, T.: Single-shell carbon nanotubes of 1-nm diameter. Nature. **363**, 603–605 (1993)
7. Alivisatos, A.P.: Semiconductor clusters, nanocrystals, and quantum dots. Science. **271**, 933–937 (1996)
8. Dresselhaus, M.S., Dresselhaus, G.: Intercalation compounds of graphite. Adv. Phys. **51**, 1–186 (2002)
9. Zhang, Y., Small, J.P., Pontius, W.V., Kim, P.: Fabrication and electric-field-dependent transport measurements of mesoscopic graphite devices. Appl. Phys. Lett. **86**, 073104 (2005)
10. Novoselov, K.S., et al.: Electric field effect in atomically thin carbon films. Science. **306**, 666–669 (2004)
11. Novoselov, K.S., et al.: Two-dimensional atomic crystals. Proc. Natl. Acad. Sci. U. S. A. **102**, 10451–10453 (2005)
12. Xia, F., Wang, H., Xiao, D., Dubey, M., Ramasubramaniam, A.: Two-dimensional material nanophotonics. Nat. Photonics. **8**, 899–907 (2014)
13. Splendiani, A., et al.: Emerging photoluminescence in monolayer MoS$_2$. Nano Lett. **10**, 1271–1275 (2010)

14. Eda, G., Maier, S.A.: Two-dimensional crystals: managing light for optoelectronics. ACS Nano. **7**, 5660–5665 (2013)
15. Bhimanapati, G.R., et al.: Recent advances in two-dimensional materials beyond graphene. ACS Nano. **9**, 11509–11539 (2015)
16. Ashton, M., Paul, J., Sinnott, S.B., Hennig, R.G.: Topology-scaling identification of layered solids and stable exfoliated 2D materials. Phys. Rev. Lett. **118**, 106101 (2017)
17. Revard, B.C., Tipton, W.W., Yesypenko, A., Hennig, R.G.: Grand-canonical evolutionary algorithm for the prediction of two-dimensional materials. Phys. Rev. B. **93**, 054117 (2016)
18. Geim, A.K., Novoselov, K.S.: The rise of graphene. Nat. Mater. **6**, 183–191 (2007)
19. Oshima, C., Nagashima, A.: Ultra-thin epitaxial films of graphite and hexagonal boron nitride on solid surfaces. J. Phys. Condens. Matter. **9**, 1–20 (1997)
20. Katsnelson, M.I.: Graphene: carbon in two dimensions. Mater. Today. **10**, 20–27 (2007)
21. Wang, L., et al.: One-dimensional electrical contact to a two-dimensional material. Science. **342**, 614–617 (2013)
22. van Wees, B.J., et al.: Quantized conductance of point contacts in a two-dimensional electron gas. Phys. Rev. Lett. **60**, 848–850 (1988)
23. Fiori, G., et al.: Electronics based on two-dimensional materials. Nat. Nanotechnol. **9**, 768–779 (2014)
24. Yazyev, O.V., Kis, A.: MoS_2 and semiconductors in the flatland. Mater. Today. **18**, 20–30 (2015)
25. Son, Y.-W., Cohen, M.L., Louie, S.G.: Energy gaps in graphene nanoribbons. Phys. Rev. Lett. **97**, 216803 (2006)
26. Choi, W., et al.: Recent development of two-dimensional transition metal dichalcogenides and their applications. Mater. Today. **20**, 116–130 (2017)
27. Kroemer, H.: Theory of a wide-gap emitter for transistors. Proc. IRE. **45**, 1535–1537 (1957)
28. Kroemer, H.: Heterostructure bipolar transistors and integrated circuits. Proc. IEEE. **70**, 13–25 (1982)
29. Sze, S.M., Kwok, K.N.: Physics of Semiconductor Devices. Wiley, New York (2006)
30. Geim, A.K., Grigorieva, I.V.: Van der Waals heterostructures. Nature. **499**, 419–425 (2013)
31. Dean, C.R., et al.: Boron nitride substrates for high-quality graphene electronics. Nat. Nanotechnol. **5**, 722–726 (2010)
32. Wang, H., et al.: Two-dimensional heterostructures: fabrication, characterization, and application. Nanoscale. **6**, 12250–12272 (2014)
33. Akinwande, D., Petrone, N., Hone, J.: Two-dimensional flexible nanoelectronics. Nat. Commun. **5**, 5678 (2014)
34. Das, S., Robinson, J.A., Dubey, M., Terrones, H., Terrones, M.: Beyond graphene: progress in novel two-dimensional materials and van der Waals solids. Annu. Rev. Mater. Res. **45**, 1–27 (2015)
35. Bonaccorso, F., et al.: Graphene, related two-dimensional crystals, and hybrid systems for energy conversion and storage. Science. **347**, 1246501 (2015)
36. Lotsch, B.V.: Vertical 2D Heterostructures. Annu. Rev. Mater. Res. **45**, 85–109 (2015)
37. Zhang, K., Lin, Y.-C., Robinson, J.A.: Semiconductors and Semimetals. **95**, 189–219 (2016)
38. Koma, A.: Van der Waals epitaxy for highly lattice-mismatched systems. J. Cryst. Growth. **201–202**, 236–241 (1999)
39. Schlom, D.G., Chen, L.-Q., Pan, X., Schmehl, A., Zurbuchen, M.A.: A thin film approach to engineering functionality into oxides. J. Am. Ceram. Soc. **91**, 2429–2454 (2008)
40. Lee, G.-H., et al.: Electron tunneling through atomically flat and ultrathin hexagonal boron nitride. Appl. Phys. Lett. **99**, 243114 (2011)
41. Weitz, R.T., Yacoby, A.N.: Graphene rests easy. Nat. Nanotechnol. **5**, 699–700 (2010)
42. Yankowitz, M., et al.: Emergence of superlattice Dirac points in graphene on hexagonal boron nitride. Nat. Phys. **8**, 382–386 (2012)
43. Britnell, L., et al.: Field-effect tunneling transistor based on vertical graphene heterostructures. Science. **335**, 947–950 (2012)

44. Lim, H., Yoon, S.I., Kim, G., Jang, A.-R., Shin, H.S.: Stacking of two-dimensional materials in lateral and vertical directions. Chem. Mater. **26**, 4891–4903 (2014)
45. Georgiou, T., et al.: Vertical field-effect transistor based on graphene-WS_2 heterostructures for flexible and transparent electronics. Nat. Nanotechnol. **8**, 100–103 (2013)
46. Withers, F., et al.: Light-emitting diodes by band-structure engineering in van der Waals heterostructures. Nat. Mater. **14**, 301–306 (2015)
47. Withers, F., et al.: WSe_2 light-emitting Tunneling transistors with enhanced brightness at room temperature. Nano Lett. **15**, 8223–8228 (2015)
48. Terrones, H., López-Urías, F., Terrones, M.: Novel hetero-layered materials with tunable direct band gaps by sandwiching different metal disulfides and diselenides. Sci. Rep. **3**, 1549 (2013)
49. Lv, R., et al.: Transition metal dichalcogenides and beyond: synthesis, properties, and applications of single- and few-layer nanosheets. Acc. Chem. Res. **48**, 56–64 (2015)
50. Fang, H., et al.: Strong interlayer coupling in van der Waals heterostructures built from single-layer chalcogenides. Proc. Natl. Acad. Sci. U. S. A. **111**, 6198–6202 (2014)
51. Chiu, M.-H., et al.: Spectroscopic signatures for interlayer coupling in MoS_2-WSe_2 van der Waals stacking. ACS Nano. **8**, 9649–9656 (2014)
52. Rivera, P., et al.: Observation of long-lived interlayer excitons in monolayer $MoSe_2$-WSe_2 heterostructures. Nat. Commun. **6**, 6242 (2015)
53. Haigh, S.J., et al.: Cross-sectional imaging of individual layers and buried interfaces of graphene-based heterostructures and superlattices. Nat. Mater. **11**, 764–767 (2012)
54. Robinson, J.A.: Growing vertical in the flatland. ACS Nano. **10**, 42–45 (2016)

Chapter 2
Synthesis and Properties of 2D Semiconductors

2.1 Introduction

In the previous chapter, brief history of the development and fundamentals of 2D materials and vdW heterostructures and the very first methods to isolate them are provided. Graphene can be considered as the funding layer for the field of 2D materials. And we are able to continuously branch out from graphene to other kinds of 2D layers, which sometimes is so called "beyond graphene" 2D layers, and also the sciences and engineering behind them. A heterostructure made of 2D semiconducting materials is an important remark toward flexible and low-power optoelectronics in the future. Analogously, 2D TMDCs represent a new class of building blocks. By combining certain of them, interesting physical sciences and practical applications can be created out of our hands. However, current methods for making a vdW heterostructure may not always provide good material interfaces. This challenge inspired my graduate research on synthetic 2D layers and their heterostructures and discovery of their properties. This chapter covers some practical aspects of thin-film deposition and also methods used for depositing 2D TMDC domains and films. The transport mechanism for 2D material devices is dominated by a few scattering events, which a lot of time are related to the interface of 2D materials and their substrates. This chapter, therefore, provides all necessary knowledges that are not all included in the later chapter which focused on the properties, devices of synthetic 2D layers, 2D/2D vdW heterostructures, and 2D/3D heterostructures.

© Springer Nature Switzerland AG 2018
Y. -C. Lin, *Properties of Synthetic Two-Dimensional Materials and Heterostructures*, Springer Theses,
https://doi.org/10.1007/978-3-030-00332-6_2

2.2 Molecular Absorption and Desorption Process During Thin-Film Deposition

The thin-film process sequence contains six substeps. First, the deposited atoms and molecules adsorb on the surface. Second, they often diffuse certain distance and then incorporate into the film. Third, in the incorporating process, the adsorbed species reacted with each other and also with the surface to form the film materials. Fourth, the initial cluster of the film materials is called nucleation. Fifth, when the film grows thicker, it establishes a structure that includes roughness and crystallography. And, sixth, diffusional interactions occur within the bulk of the film or with the substrate.

When a molecule approaches the surface within a few atomic distances, it will start to feel an attraction by interacting with the surface molecules. It happens because the molecules and atoms act as oscillating dipoles, and this behavior induces dipole interaction known as van der Waals force/London dispersion force. If the molecule is a polar one and has permanent dipole, the attraction is stronger. This molecule is trapped in a weakly adsorbed state in the beginning called physical adsorption (physisorption). The fraction of approaching molecules so adsorbed is called the trapping probability (δ). Intuitively, the fraction of the molecules that reflect or escape is $1 - \delta$. Generally, the substrate is at an elevated temperature and is thermally accommodated to the molecules during the deposition process. This thermal energy makes the physisorbed molecules mobile, so they will diffuse between surface atomic sites. After a while, it may either desorb by gaining enough energy or undergo a further interaction including the formation of chemical bonds with the surface atoms, that is, chemisorption. Chemisorption involves the electron sharing in new molecular orbitals and is much stronger than physisorption, since the later only involves dipole interactions. Not all of the vapors would trap and condense on a foreign substrate. The physisorbed molecules will eventually escape the substrate before they become chemisorbed ones. Thus, the chemisorption reaction probability, η, is defined as the fraction of the arriving vapor that becomes chemisorbed on a foreign substrate.

Some of the physisorbed species eventually escape back into the vapor phase; the sticking coefficient, S_c, is used to denote the fraction of the arriving vapor that remains adsorbed for the entire duration of the experiment. S_c is very useful in thin-film deposition, since it's equal to the fraction of arriving that becomes "incorporated" as part of the film. The incorporation means this arriving vapor becomes adsorbed and subsequently buried before it can desorb. The precursor adsorption can also be simply described by a diagram of the potential energy versus molecular distance from the surface (z). The potential energy is commonly expressed as the molar quantities (E_p). One curve shown is for the precursor state and another one is for the chemisorbed state in the Fig. 2.1. The tails of the two curves will intersect at a certain distance from the surface forming an energy barrier, E_a, which is an activation energy that the arriving vapor needs to overcome in order to become dissociatively chemisorbed.

Typically, the E_p of the element involved in deposition processes is set to zero as it is in thermodynamic standard state ($Y_{2(g)}$ in this case, lining at zero E_p). One main

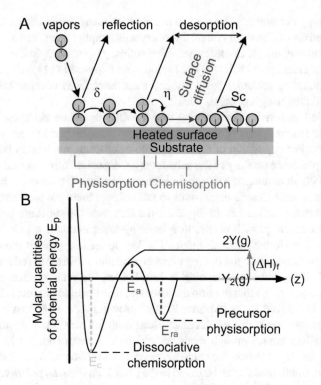

Fig. 2.1 (**a**) Adsorption processes and important quantities. (**b**) Energetics of the precursor adsorption model

advantage of the energy-enhanced deposition processes (i.e., perform deposition at higher temperature) is that the arriving molecules can conquer the E_a barrier. There are two ways in which arriving vapor can have $E_p > 0$ at the surface, either as high kinetic energy of accelerating molecules or high potential energy of dissociated ones (i.e., formation energy provided to $2Y_{(g)}$). Gases have their E_p raised by being dissociated. Solids and liquids have theirs raised by being evaporated. If the E_p of the arriving vapor is high enough, direct chemisorption can happen without going through the precursor state. That is to say, the atoms and molecules of the arriving vapor instantly react with the surface and then make film deposition (For more related knowledge, I recommend further reading on *Thin-Film Deposition: Principle & Practice* by Donald L. Smith) [1].

2.2.1   Nucleation and Growth

The fundamental concept for nucleation behavior is surface energy, which is the work energy stored in a new surface after it was created. For solids, surface energy tends to minimize itself by surface diffusion. This process subsequently determines the structure of thin films. In thin-film growth, area of surface topography and surface energy per unit area (γ) vary in accord with many properties of the exposed surface in terms of chemical composition, crystallographic facet, and atomic reconstruction and roughness. In most crystalline solids, γ is anisotropic and only one or two of many facets provide low γ. For example, an exposed {111} face of Si and Ge that have diamond structure has a lower γ than other faces because this face has fewer unsatisfied dangling bonds sticking out.

In layered materials like graphite and TMDC, there are no chemical bonds between the atomic layers of the basal plane, and thus the basal plane is their low-energy facet. For deposition of thin films onto a substrate, nucleation behavior has a strong dependence on the γ of the substrate (γ_r), deposited film (γ_f), and their interface (γ_i). With an assumption that there is enough surface diffusion so that depositing materials can rearrange themselves to minimize γ, there are two situations on a bare substrate for nucleation. In Fig. 2.2a, the film wets the substrate because "$\gamma_f + \gamma_i < \gamma_r$," so that the growth occurs in a layer-by-layer manner, which is so called "Frank-van der Merwe" growth mode. The key to let this growth mode occur is there must be strong enough bonding between film and substrate to reduce γ_i. On the other hand, if the substrate bonding is insufficient, the total surface energy will become "$\gamma_f + \gamma_i = \gamma_r$," so that the film does not wet the substrate but forms 3D islands. This mode is referred to the "Volmer-Weber" growth mode, as shown in Fig. 2.2b. The third growth mode that often comes along with the previously mentioned ones is "Stranski-Krastanov" growth mode, in which the growth changes from layer to island after one layer or two due to changing energy situation with successive monolayers (more fundamentals can be found in the book *Thin-Film Deposition: Principle & Practice* by Donald L. Smith) [1].

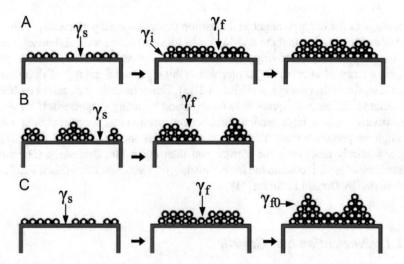

Fig. 2.2 Film growth modes: (**a**) Frank-Van der Merwe (layer), (**b**) Volmer-Weber (island), and (**c**) Stranski-Krastanov (layer+island) [2]

2.2.2   *Epitaxial Relationship Between Deposited Materials and Substrates*

Epitaxy means a crystalline overlayer deposited on a crystalline substrate. The crystallographic order of the deposited film is significantly influenced by the crystallinity of the substrate, thus achieving certain degree of matching between the two along the interface. Crystal symmetry is one of the fundamental criteria for epitaxy. If it is interrupted, its potential energy (E_p) increases because the angle and length of bonds and the number of bonds attached to an atom of the crystal change. Consequently, this interruption introduces excess energy to surface and interface per unit area (γ). In order to minimize the γ when one crystal is deposited on a crystalline substrate, the density of bonds of appropriate length and angles needs to be maximized in an attempt to merge symmetries between themselves. The way that the deposited material minimizes the γ is to crystallographically align itself with the substrate as to match the substrate's bonding symmetry and crystal periodicity, in another word to grow epitaxially. From a synthesis point of view, acheiveing a successful epitaxy requires 1) the substrate symmetry will not be screened by any interfacial disorder, and 2) the growth temperature is high enough so that the depositing atoms can rearrange themselves into equilibrium position before incorporating into the film. In general, epitaxy can be either homotype or hetero-type – the former type is for the growth of material onto itself, whereas the latter is for the one on other substrates that results in $\gamma > 0$. The preferred crystallographic orientation of the heteroepitaxial film is often which γ can be minimized. One fundamental criterion for epitaxy is relatively small lattice (frictional) mismatch in the atomic periodicities of the material/substrate along the interface, which is defined as:

$$f = \frac{(\alpha_e - \alpha_s)}{(\alpha_e + \alpha_s)/2},$$

where α_e and α_s are the atomic spacings along one particular crystallographic direction in the film crystal and in the substrate surface, respectively [2]. Despite f can change at different growth temperature due to the difference in the thermal-expansion coefficients of the film and substrate, the room-temperature value is the generally discussed. If f is too high (>0.1), only a few interfacial bonds are aligned well that γ cannot be minimized. The option of good single-crystal substrates for epitaxy is limited because it is not easy to find a large-area substrate that also has low defect density. They are also required to be chemically robust or damage-resistant. Some of the commercially available that have reasonable quality, size, and cost include Ge, GaAs, sapphire, mica, and SiC.

   In order to achieve an ideal epitaxial film with atomically sharp interface, one must consider chemical compatibility in reaction, a deposition process that is not operating near equilibrium, whereby the incorporation flux of adsorbed vapor into the film is larger than reevaporation flux of film material and also a small lattice mismatch. The attractive combinations for device applications of heteroepitaxy are

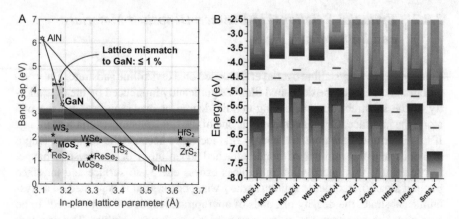

Fig. 2.3 (**a**) Bandgap versus in-plane lattice parameter for III-nitrides and TMDC. (**b**) Band alignment of TMDC in hexagonal (H) and trigonal (T) phase [3, 4]

those that obtain large bandgap difference and low f simultaneously. For example, heteroepitaxial films integrating group III nitride (N) semiconductors including GaN, AlN, InN, and their alloys are common nowadays for compact, energy-efficient, light-emitting diodes as well as for high-power electronic devices. Heteroepitaxial films made of group III-N substrates and 2D TMDC (e.g., WSe_2/GaN) are getting more popular because their mutually small lattice mismatch and the possibility to create a large band-edge offset can result in high-performance semiconductor devices with high-quality interface (Fig. 2.3a) [3]. Similarly, a variety of semiconducting TMDC with different bandgap size and position can also create high-quality heteroepitaxy with a significantly large band offset well suited for optics and electronics (Fig. 2.3b) [4] (This section is referenced to the book *Thin-Film Deposition: Principle & Practice* by Donald L. Smith.) [1].

2.3 Synthesis Techniques for 2D TMDC

Synthesis of bulk TMDCs has been explored for many years and already had many routes (Fig. 2.4) [5]. For example, chemical vapor transport has been used to synthesize a variety of TMDC under equilibrium conditions using a transport agent (B_2 or I_2) to transport transition metals and chalcogen atoms across a thermal gradient in a vacuum-sealed ampule. Despite this, the process requires days and weeks; the resulted bulk crystallites provide ultrahigh quality for researchers. Similarly, direct vapor transport utilizes a thermal gradient to vaporize stoichiometric TMDCs (many times in powder form) and to recrystallize them at the cold end of the furnace. Although this route has been successful for production of a wide variety of materials (MoS_2, WS_2, MoS_2, WSe_2, $TaSe_2$, etc.) that can be further mechanically exfoliated to monolayers, it is not scalable and thus cannot fulfill many applications that require large-area samples.

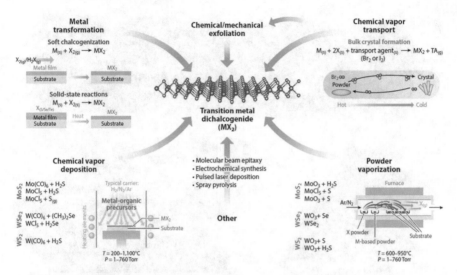

Fig. 2.4 Summary of primary growth techniques for the formation of TMDC atomic layers. These methods include chemical vapor deposition, powder vaporization, metal transformation, chemical vapor transport, chemical exfoliation, pulsed laser deposition, molecular beam epitaxy, spray pyrolysis, and electrochemical synthesis [5]

In order to increase the area size and uniformity of film thickness for synthetic TMDC thin films, researchers came out the chalcogenization process [6], in which thin films made of transition metal/transition metal oxide were converted into MX_2 after exposure to chalcogen vapor, such as $S_{(g)}$, $Se_{(g)}$, and $H_2Se_{(g)}$. Despite this process indeed provides excellent uniformity along both of lateral and vertical direction, its nanoscale domain size and nearly amorphous nature are the main detriment to high-performance optoelectronics.

Current state-of-the-art techniques for high-quality monolayers are powder vaporization (PV) and metal-organic chemical vapor deposition. Both of these two methods have demonstrated large domain (edge length > 100 μm) and wafer-scale-size films deposited on insulating substrates. Therefore, they will be further discussed and implanted in this thesis.

2.3.1 Powder Vaporization

The vapor-phase reaction or powder vaporization (PV) was developed for vapor-phase growth of crystalline MoS_2 monolayer on SiO_2 in the first paper of synthetic monolayer [7]. This technique provides the easiest method for scalable deposition of high-quality TMDC films on any arbitrary substrate (Fig. 2.5a). Taking MoS_2 as an example, sulfur and MoO_3 powders were chosen as the precursors because they can be vaporized easily at low elevated temperature. The Mo-O-S ternary phase

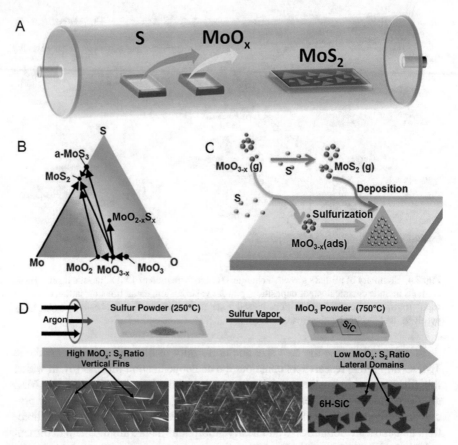

Fig. 2.5 (**a**) Schematic illustration of commonly PV method for TMDCs monolayers. (**b**) Phase diagram and possible reaction routes for MoS$_2$ growth. (**c**) Schematic illustration of the gas-phase reaction and surface epitaxy of MoS$_2$. (**d**) During PV of MoS$_2$, a transition from vertical domains to a mixture of vertical and horizontal domains and finally only horizontal domains, as the partial pressure ratio of MoO$_x$:S$_2$ decreases toward the end of the reactor [8, 9]

diagram in Fig. 2.5b indicates that the gas-phase MoO$_3$ precursors may undergo a two-step reaction during the growth [8]:

$$MoO_3 + (x/2)S \rightarrow MoO_{3-x} + (x/2)SO_2, \text{ and then } MoO_{3-x} + (7-x/2)S \rightarrow MoS_2 + (3-x/2)SO_2$$

The transition metal sub-stoichiometric oxides are also formed during the reaction. As illustrated in Fig. 2.5c, the intermediated adsorbates diffuse to the substrate surface and further react with sulfur vapors to grow MoS$_2$ layers. MoS$_2$ clusters may also form before it lands on the surface and becomes adsorbate. The partial pressure (which is dictated by temperature) of S and MoO$_3$ governs subsequent adsorption on the substrate and film morphology when they are traveling toward the downstream. An investigation performed by Vila et al. shows a high MoO$_x$:S$_2$ partial pressure near the front of the substrate which promotes MoO$_2$ growth, whereas MoS$_2$ monolayers grow near the end of the substrate, whereby the partial pressure is lower. Besides, the excess in sulfur supply will suppress the volatilization of MoO_3, make the partial pressure of vaporized MoO_x low, and thus make domain size small. Therefore, a more controllable vapor pressure is in demand in order to have a more consistent morphology and quality.

2.3.2 Metal-Organic Chemical Vapor Deposition

To ensure a consistent precursor supply and improve the scalability and controllability for TMDC deposition, metal-organic chemical vapor deposition (MOCVD) was developed. It uses organic compounds that contain transition metal and chalcogen elements as the precursors for synthesis. The system for MOCVD process can be hot-wall and cold-wall type (Fig. 2.6) [5, 10]. The precursors with a high equilibrium vapor pressure are required for MOCVD process so that they can be delivered through mass flow controller. Molybdenum/tungsten hexacarbonyl ($Mo(CO)_6$/$W(CO)_6$) and dimethyl/diethyl-sulfide/selenide (($CH_3)_2S$, ($C_2H_5)_2S$, ($CH_3)_2Se$, ($C_2H_5)_2Se$) are common options for making TMDC in MOCVD. By using mass flow controller and controlling vapor pressure of each precursor with a bubbler, MOCVD process has been proven to have better control than the PV and other chalcogenization, in terms of ratio of partial pressure and flow rate and deposition rate.

One concern on the MOCVD process for TMDC is unintentional carbon incorporation. Chalcogen source such as ($CH_3)_2Se$ will crack at high temperature in the growth and unavoidably deposit carbon thin layers on the substrate surface and interrupt the film morphology [11, 12]. Due to this reason, the alternative precursor like carbon-free H_2Se and H_2S has gradually been adopted. A comparison between using H_2Se and DMSe by Zhang et al. shows that the carbon incorporation has been removed and the film morphology has also been improved in Raman spectrum and AFM topography (Fig. 2.7) [12]. Despite the carbon-contained precursors can diminish the quality of growth on sapphire, it may not severely affect the growth of TMDC on other substrates.

2.3.3 Epitaxial Graphene Synthesis

In order to provide optoelectronic applications with uniform and large-scale graphene, the synthesis of epitaxial graphene (EG) on silicon carbide (SiC) has been developed [13]. SiC wafer can be fabricated in the range of 2–6 inches in diameter using standard industrial semiconductor synthesis techniques. The very first growth of graphene on SiC was performed in ultrahigh vacuum demonstrated by Van Bommel et al. in 1975 [14]. Silicon sublimation from the SiC causes a carbon-rich surface that provides nuclei for graphene growth. However, the electrical transport of epitaxial graphene by UHV method did not look great mainly because of high Si sublimation rate, which results in a poor topography. Hence, it is necessary to lower

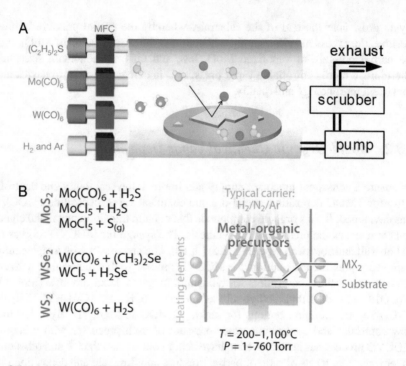

Fig. 2.6 (**a**) A hot-wall MOCVD reactor and (**b**) a cold-wall reactor that use induction heating and graphite susceptor during the growth [5, 10]

the rate at which silicon sublimes. Among many proposed methods for controllable Si sublimation, one promising route is "confinement controlled sublimation" [15], which encloses SiC in a graphite crucible during silicon sublimation (Fig. 2.8a). The SiC substrate is first cleaned via chemical solutions and then H-etched at 1500 °C in 700 Torr of H$_2$/Ar, which removes polishing damages and results in a surface with atomically flat terraces. The graphene is then obtained on a SiC substrate via the solid-state decomposition of the substrate, which is achieved by annealing the material in elevated temperatures in the ranges of 1600–2000 °C in partial pressures of Ar, driving the sublimation of Si atoms from the surface slowly [16]. The C atoms left behind would reorganize themselves in a hexagonal fashion forming graphene [16]. By optimizing the synthesis conditions of EG, mono- to few-layer graphene deposited on the wide terraces of SiC, separated by a few unit cell high of SiC, and the conjunctions of SiC step/terrace, respectively (Fig. 2.9) [17].

2.4 Vertical and Radical Heterostructures Based on Synthetic 2D Materials

The practically useful heterostructures made of III–V compounds, such as heterostructure bipolar transistors, phototransistor with wide-gap emitters, and double-heterostructure lasers, hadn't appeared until the growth technologies of MOCVD

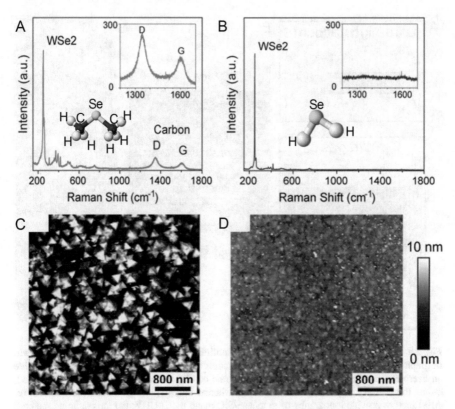

Fig. 2.7 Raman spectra of WSe_2 grown using (**a**) DMSe and (**b**) H_2Se. The insets are zoom-in regions showing the D and G peaks of carbon in WSe_2 grown with DMSe and the AFM topography of WSe_2 grown using (**c**) DMSe and (**d**) H_2Se [12]

and MBE were developed in the early 1970s [18]. Similar to the early development on their conventional counterparts, vdW heterostructures haven't been practical since its demonstration in 2010 due to limited size of clean interface obtained and the absence of techniques for the large-area growth. Although the vdW epitaxy, growing one vdW solid on another, have already been recognized in the 1980s, many were by Koma [19], and it did not get much attention from the research societies until the breakthrough results exploited in manually stacked vdW heterostructures. Recently, the emergence of direct synthesis of vdW solids, utilizing CVD, MOCVD, and MBE [5], also made impressive progress in synthetic vdW heterostructures like graphene-hBN transistors [20], graphene-TMDC photosensors [17, 21], and TMDC $p–n$ junctions and tunneling diodes grown on graphene [22, 23] and insulating substrates [24, 25]. In view of these recent results, synthetic vdW heterostructures appear to revolutionize the digital electronics and their industries.

In order to synthesize crystalline TMDC layers, lattice of the selected substrate is critical for the epi-growth of vdW heterostructures. Shi et al. [26] initiatively used CVD graphene grown on copper foils as the template for MoS_2 growth (Fig. 2.10a).

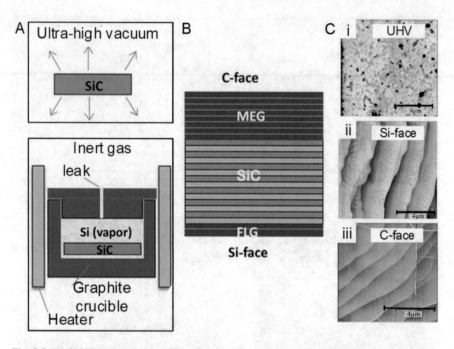

Fig. 2.8 (**a**) If SiC is annealed in UHV, silicon sublimation is not confined, causing rapid growth of graphene. The confinement controlled sublimation method uses a graphite crucible to provide an overpressure of Si vapor so that the growth rate can be controllable. (**b**) Using this method, mono- to few-layer graphene grows on Si-face, whereas thin graphite grows on C-face of SiC. (**c**) AFM images provide topography of graphene/SiC made by (i) UHV; (ii) the confining method: Si-face; and (iii) C-face [15]

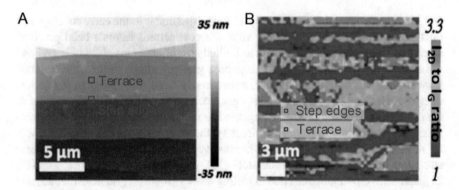

Fig. 2.9 (**a**) Topography of graphene/SiC cannot identify the layer number, which can be revealed by the Raman spectroscopes. (**b**) The ratio of the intensity of graphene 2D to G peaks ($I_{2D/G}$) can identify graphene ($I_{2D/G} \geq 2$) and the few layers ($I_{2D/G} \leq 1$) [17]

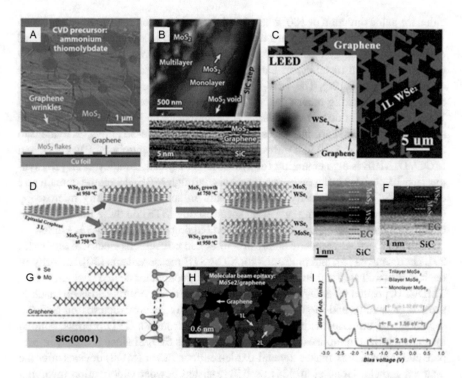

Fig. 2.10 (a) Multilayer MoS_2 grown on CVD graphene/Cu through the thermal decomposition of ammonium thiomolybdate. (b) Mono- and few-layer MoS_2 grown on epitaxial graphene (EG) from powder vaporization (PV) process. (c) WSe_2 monolayers grown EG through PV process or metallic-organic CVD (MOCVD). The WSe_2 lattices are fully registered to the graphene lattices, as evident by low-energy electron diffraction pattern (LEED, Inset). (d) The flow for growing "trilayer" vdW heterostructures. MoS_2-EG was converted into $MoSe_2$-EG during the growth of WSe_2 layers via a Se-S ionic exchange occurring in high temperatures. (e–f) STEM image confirms these trilayer stacks exhibit pristine interfaces without W-Mo or S-Se formation. (g–i) $MoSe_2$ layers ranging from 1 L to 3 L grown on bilayer EG by MBE. EG serves as bottom electrodes for STS in (i), which measures the quasiparticle bandgap of 1 L–3 L $MoSe_2$ [27]

The reported process utilizes $(NH_4)_2MoS_4$ precursors that were thermally decomposed into MoS_2 in vapor phase and then subsequently deposited on CVD graphene/Cu foil. The as-grown MoS_2 domains on graphene adopted the same orientation of underlying graphene. This experiment indicated that an epitaxial vdW heterostructure can be realized still; even the lattice mismatch can be 20–23% [5, 26]. Similarly, the study in this thesis used epitaxial graphene (EG)/SiC as the growth template for monolayer MoS_2 made via powder vaporization (Fig. 2.10b). We also found morphology and defects of EG/SiC can significantly impact the nucleation density and thickness of MoS_2 layers [17]. Scanning transmission electron microscopy (STEM) images show that the atomically sharp interface is possible to achieve through vapor deposition techniques (Bottom, Fig. 2.10b). In addition, it is also possible to grow larger domain of WSe_2 monolayers on EG/SiC via vdW epitaxy [22]. Following vdW epitaxy, monolayered WSe_2 domains grown on graphene consistently align at either the same direction or 180° rotated and thus achieve commensurability between WSe$_2$ and graphene (scanning electron microscopy (SEM) image in Fig. 2.10c), as evident by low-energy electron diffraction (LEED) patterns, which roughly show that four transition metal atoms can align with nine carbon atoms in a long-range order (Inset, Fig. 2.10c). The vdW heterostructures can be put on a more sophisticated level by stacking other types of TMDs layers. Two-step growth of MoS$_2$ and WSe$_2$ was carried out to create MoS$_2$-WSe$_2$-graphene and WSe$_2$-MoSe$_2$-graphene that have clean and sharp interfaces without Mo-W and Se-Se alloys, as evident in STEM images (Fig. 2.10, d–f). Besides the techniques of pyrolysis, PV, CVD, and MOCVD, MBE is also emerging for synthetic 2D crystals. Bradley et al. [28] synthesized 1 L to 3 L MoSe$_2$ on bilayer graphene through MBE. Albeit the domain size of the MoSe$_2$ being typically less than 1 µm in scanning tunneling microscopy (STM), performing scanning tunneling spectroscopy (STS) on these films is able to obtain the quasiparticle bandgaps and exciton binding energy of 1 L–3 L MoSe$_2$.

Although TMDC-graphene heterostructures can be useful as electrical diodes [5, 29], photosensors [5], and platforms for STM/STS measurements [30], majority of devices research are focusing on metal-oxide-semiconductor device geometry (e.g., TMDCs deposited on SiO$_2$/Si). To fulfill this need, many efforts had been made to grow high-quality and large-size TMDC-based vdW heterostructures on SiO$_2$/Si, sapphire, and other insulating substrates, mainly through a CVD process. Among insulating growth templates, the most popular one is SiO$_2$/Si since it is easy to prepare and immediately makes a metal-oxide-semiconductor (MOS) devices after the material growth. Gong et al. [24] used Te-assisted powder vaporization involving the reaction of MoO$_3$, W, and S powders to grow both of the lateral and vertical MoS$_2$-WS$_2$ heterostructures in an in situ process. The role of Te powders involved is for lowering the melting point of W powders via forming metastable Te-W alloys during the reaction [24]. The lateral MoS$_2$-WS$_2$ grows at 650 °C (Fig. 2.11a,b), while the vertical one grows at a higher temperature, at 850 °C (Fig. 2.11c,d) [24]. Besides the heterostructure using single chalcogen atom, Li et al. [25] also developed a two-step ex situ process using the edges of the WSe$_2$ monolayers pre-grown at 950 °C as nucleation sites and then growing MoS$_2$ monolayers epitaxially around the MoS$_2$ monolayers at 700 °C to obtain MoS$_2$-WSe$_2$ lateral heterostructures (Fig. 2.11e). The order for material growth, that is, WSe$_2$ first and MoS$_2$ second, is deliberately decided to avoid the ionic exchange of Se-S occurring above 800 °C. The STEM performed on MoS$_2$-WSe$_2$ confirmed that the lateral interface is atomically abrupt and no sign of Mo-W and Se-S formation in a micrometer range in parallel to the junction (Fig. 2.11f–h) [25]. Besides the above "flat" cases, vdW heterostructures can also exist in a vertically aligned fashion. Jung et al. [31] sulfurized (selenized) patterned Mo/W arrays to synthesize MoS$_2$-WS$_2$ (MoSe$_2$-WSe$_2$) heterostructures in large area (Fig. 2.11i). Functionality and properties of this type of structures may be completely controllable because the dimension and thickness of Mo/W arrays can be controlled by the lithography and sputtering time, respectively. Although electrical transports don't favor the vertical formation, as evident in STEM images (Fig. 2.11j) (the measured mobility is <0.01 cm^2 V^{-1} S^{-1} in both of the direction vertical and parallel to the array), preferably exposed dangling bonds on the edge sites are useful for hydrogen evolution reaction [31].

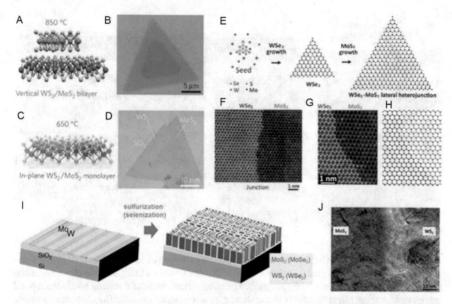

Fig. 2.11 (**a, b**) Schematic illustration and optical images of the vertical WS_2-MoS_2 heterostructures. (**c, d**) Schematic illustration and optical images of the lateral WS_2-MoS_2 heterostructures. (**e**) Schematic illustration of the growth process of the monolayer WSe_2-MoS_2 lateral heterostructure on sapphire. (**f, g**) High-resolution STEM images of the WSe_2-MoS_2 lateral heterostructure at the interface regions. (**h**) Atomic model shows the interface structure of the WSe_2-MoS_2 lateral heterostructure. (**i**) Sputtered Mo-W strips patterned on SiO_2 were converted into MoS_2-WS_2 ($MoSe_2$-WSe_2) vertically aligned heterostructures after sulfurization (selenization). (**j**) STEM shows the interface of vertically aligned MoS_2-WS_2 heterostructures. A higher Z number of W atoms makes WS_2 region brighter than the other in STEM [27]

2.5   2D Materials Electronics: Interface Is Critical

Atomically thin TMDC layers have been considered as prominent alternatives of bulk Si, Ge, and III–V compound semiconductors (Fig. 2.12a,b) for future low-power electronic technology and because their ultrathin nature leads to excellent carrier confinement effects (Fig. 2.12c,d). In addition, comparing to 3D elemental and compound semiconductors, TMDCs have no dangling bonds that could potentially lead to traps that reduce the mobility of electrons and holes in devices.

In graphene- and 2D TMDC-based transistors, transport and scattering activities are confined to their plane. Wang et al. summarized the scattering mechanisms that affected the mobility of carriers [33]: acoustic and optical phonon scattering, Coulomb scattering at charged impurities, surface interface phonon scattering, and roughness scattering. In the electronics of 3D materials, interfacial roughness scattering can dominate because they mainly rely on the quantum well structure. In the electronics of 2D materials, on the other hand, the effect of surface phonon scattering and Coulomb scattering can be very important in 2D materials electronics. 2D TMDC has partial ionic bonds between the metal and chalcogen atoms; crystal

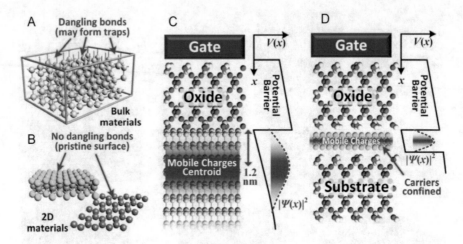

Fig. 2.12 The inherent difference in surface properties between (**a**) 3D materials and (**b**) 2D materials significantly impacts performance of the devices fabricated on them. Mobile charge distribution in (**c**) 3D and (**d**) 2D materials indicates a pristine surface without dangling bonds has a good carrier confinement effect that leads to excellent gate electrostatic control. The blue plumes inside the wells of (**c**) and (**d**) are electron density waves and $V(x)$ stands for a potential well [32]

deformation would lead to polarized fields that scatter carriers. The phonon scattering has a strong relationship with the carrier mobility and heavily depends on operating temperatures. Calculation from first principles by Kaasbjerg et al. shows the temperature dependence of carrier mobility in 1 L MoS$_2$ with a constant carrier concentration (Fig. 2.13a) [33].

Coulomb scattering in 2D semiconductors is caused by random charged impurities located within the layer or on its surfaces. This Coulomb effect also limits the mobility of graphene on SiO$_2$ to values less than 10^4 cm^2 V^{-1} s^{-1}. A 20× improvement to mobility can be achieved by placing graphene on hBN, which has much less Coulomb scattering on graphene than many substrates due to its excellent surface inertness. Engineering the dielectric environment can enhance the carrier motilities of TMDC transistors, from ~0.1–3 cm^2 V^{-1} s^{-1} to ~200 cm^2 V^{-1} s^{-1} (MoS$_2$ as an example) at room temperature [34, 35]. Screening the Coulomb scattering from the charged impurities on 2D semiconductors has been demonstrated by using the materials with a high dielectric constant [35]. The combined effect of phonons and charged impurities on the mobility of 1 L- and few-layer MoS$_2$ and other TMDC is summarized in Fig. 2.13b.

Theoretical simulation shows that 2D TMDC can make transistors with excellent performance. The operation of 1 L MoS2 field-effect transistor (FET) predicted by Yoon et al. using the non-equilibrium Green function formalism demonstrates that the top-gated MoS$_2$ transistors with gate lengths of 15 nm can reach on current as high as 1.6 mA·μm^{-1} during operation, subthreshold swing (SS) as low as 60 mV/dec, and current on/off ratio up to ten orders [34]. The illustration and the simulated

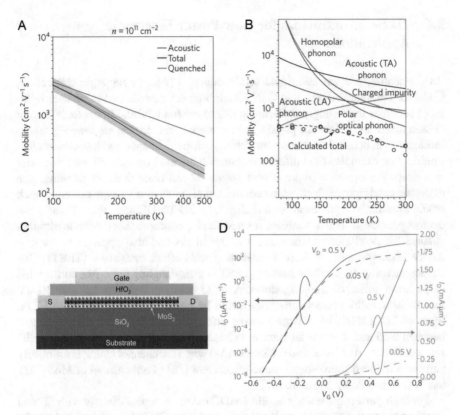

Fig. 2.13 (**a**) Carrier mobility in 1 L MoS$_2$ as a function of temperature calculated from first-principles DFT calculation for the electronic band structure, phonon dispersion, and electron-phonon interactions. The gray band shows the uncertainty in calculated mobility values due to a 10% uncertainty in computed deformation potentials associated with phonons. (**b**) Calculated and measured carrier mobility in multilayer MoS$_2$ as a function of temperature, showing the scattering contributions from charged impurities (red line), homopolar out-of-plane phonons (green lines), and polar-optical phonons (blue line), as well as the total mobility due to the combined effects (dashed line). (**c,d**) I_d-V_g characteristics at $V_d = 0.05$ V and 0.5 V on logarithmic (left axis) and linear scales (right axis). For the nominal device simulated, a maximum on current close to 1.6 mA/μm and a subthreshold swing (SS = ∂VG/∂ log10(ID)) close to 60 mV/dec are achieved. Drain-induced barrier lowering (DIBL) is as small as 10 mV/V even with very short channel length [33, 34]

transfer characteristics (gated voltage versus drain current, I_d-V_g) for a 1 L MoS$_2$ FET at different drain voltage are shown in Fig. 2.13c,d. The high-*k* dielectric simulated in this device, HfO$_2$, also improves the mobility of monolayer MoS$_2$ owing to the screening of Columbic scattering. The first experimental demonstration of a top-gated transistor based on 1 L MoS$_2$ by Kis and co-workers in 2011 has verified the theoretical simulation and showed the use of dielectric layer is effective for performance enhancement [35]. Their devices showed high on/off current ratio (~10^8), room-temperature mobility ~40 cm^2 V^{-1} s^{-1}, and SS of 60 mV/dec.

2.6 2D Semiconductors for Low-Power Electronic Applications

As the scaling task on traditional semiconductors (e.g., Si for logic electronics, GaAs for communications, and GaN for high-power devices and solid-state lightning) becomes increasingly challenging due to the fundamental limits from both of chosen materials and device physics, new materials and critical engineering breakthrough are in demand to achieve the future high-performance and low-power electronics. For example, field-effect transistors fabricated on bulk Si will not offer significant performance improvement below 22 nm node devices; thinning gate dielectric can improve electrostatic control but also increase leakage current, which would eventually degrade carrier mobility [36]. 2D TMDC and other 2D semiconductors provide several advantages for new device concepts. They provide ultrathin channel for good electrostatic control in transistors and also opportunity for new device concepts; one of these is tunneling field-effect transistors (TFET). One advantage using 2D semiconductors as FET is their dangling bond-free surface that helps lower interface trapping states. A finite bandgap between 1.5 and 2.2 eV means the standby current (off-state) would be much lower than that of Si FETs. Most of 2D TMDC have large relative effective mass for holes and electrons, between 0.55 and 0.66 for all carriers in MoX_2. This brings an advantage over 3D semiconductors (0.3 for Si and 0.15 for GaAs) when the channel length is ultimately narrow: a reduced source-drain tunneling current [36] (4 nm channel of MoS_2 FET has been achieved by Xie et al. [37]).

For high-performance devices, 2D TMDC cannot compete directly with Si- and III–V materials due to relatively low carrier mobility (TMDC, <100 cm^2 V^{-1} s^{-1}; Si, few hundreds of cm^2 V^{-1} s^{-1}; III–V, few thousands of cm^2 V^{-1} s^{-1}) [36]. However, their high on/off ratio (>10 [6], MoS_2) and low SS (<80 mV/dec, MoS_2) make them possible to outperform existing Si-based low-power devices [35]. One of leading transistor candidates for lower-power device is TFET, which can achieve low SS by creating band-to-band tunneling via field control [36]. Several theoretical and experimental works have predicted the feasibility of using 2D TMDC for TFET (Fig. 1.8a) [36], considering their excellent electrostatic control and possible low trapping states on surface. For next-generation flexible electronics, 2D materials can also be ideal candidates because they can be synthesized in polycrystalline form and in large area and are also mechanically flexible. In addition, 2D materials may have better mobility and high on/off ratio, compared to organic semiconductors (Fig. 1.8b) [36].

Although 2D materials offer tremendous opportunities to device applications, growth of large-area, high-quality, and single-crystal 2D semiconductors remains a daunting task. Processing of 2DMs is still in its infancy because the nucleation phenomena and associated surface science and defect control at the monolayer are yet understood. The community is still understanding baseline properties by using micromechanical cleaving method that provides high-quality but small size samples. The most common methods for synthetic films are chemical vapor deposition

(CVD), molecular beam epitaxy (MBE), and physical vapor processing (Fig. 1.8c) [5, 36]. Single-crystalline graphene can be grown up above 1 cm² in diameter on Cu, Cu-Ni alloys, germanium (110), and SiC via high-temperature process [36, 38]. Although current deposition techniques are able to grow single domains of 2D semiconductors with the edge size up to hundreds of μm below 1000 °C, the uniformity and crystallinity of these results are yet to meet the industrial standard for massive device fabrication. The other engineering questions, such as layer thickness control, defect density control, forming ohmic contact, and doping engineering, also need to be answered before high-performance electronics can be made on synthetic 2D materials.

2.6.1   Scalable Process for Synthetic 2D Semiconductors

It is necessary to grow highly crystalline semiconducting TMDC not only atomically thin but also laterally scalable (i.e., from sub-mm to inch-size) to make them compatible with the manufacture process in the electronic industries, which usually use inch-size wafers for mass device fabrication. In the past decade, the research communities of 2D materials (not including graphene and hBN here) have been very productive on scaling up the area of synthetic 2D TMDC using processes like chalcogenization (which is sulfurizing/selenizing transition metal/oxide thin films), MOCVD and ALD, and PV (powder vaporization using powder precursors). These methods frequently utilize crystalline wafers, like sapphire (Fig. 2.14a), SiC, and mica, which provide large-area template, good thermal and chemical stability, and also ideal crystal facets to obtain good crystallinity on the synthetic films [6, 39]. These substrates also allow us to peel off as-grown 2D TMDC films and transfer them to other surfaces, such as polymer flexible substrates and other kinds of substrates that are unsuitable for the growth (Fig. 2.14b). MOCVD and ALD processes do not necessarily require a crystalline substrate. They can directly deposit a uniform, polycrystalline TMDC thin films on SiO_2 and SiN_x for massive complimentary metal-oxide-semiconductor production (Fig. 2.14c) [10]. MOCVD/ALD process can grow TMDC on thin oxide layers at low temperatures. And, therefore, it can avoid degrading the oxide layers. In addition, their large-area films can be easily peeled off and then stacked together into vdW heterostructures (Fig. 2.14d,e) [40].

While elemental impurities present in materials growth could degrade the electrical performance of synthetic TMDC films, they sometimes can actually facilitate large domains and wafer-size coverage. This exception is evident by increasing use of elemental sodium or halide compounds, such as NaCl, KCl, or KI, in both of PV and MOCVD processes [10, 41]. For instance, using glass as a substrate in the growth can lead to 1–5 mm edge length of a monolayer domain once Na is released from the glass substrate and participates the crystal growth (Fig. 2.14f,g) [42]. Similarly, the same method utilizing Na as a catalyst has been demonstrated for the synthesis of cm-size 2D MoS_2 films on a 6-inch glass substrate slightly above 700 °C [43]. Moreover, this work provides an unprecedent uniformity on monolayer

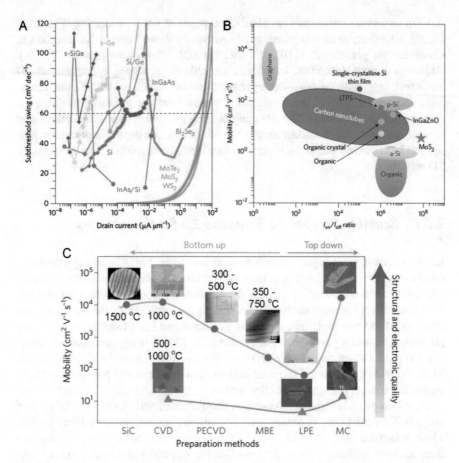

Fig. 2.14 (**a**) Measured subthreshold swing (SS) as a function of drain current for experimental tunnel field-effect transistors versus simulations. The dashed line refers to the lowest SS, 60 mV/dec, achievable in thermionic devices. (**b**) Trade-off between field-effect mobility and on/off ratio for materials typically used in flexible electronics versus organic materials. (**c**) Mobility of 2DMs as a function of preparation method. Images from left to right: low-energy electron microscopy image of epitaxial graphene on SIC, large graphene domains grown on Cu, high-resolution transmission electron microscopy (HRTEM) image showing graphene grown by plasma-enhanced CVD (PECVD), scanning transmission microscopy image of graphene grown by MBE, HRTEM image of graphene prepared by liquid-phase exfoliation, and an atomic force microscopy (AFM) image of graphene prepared by micromechanical cleavage (MC). The red curve shows the mobility of graphene on SiO_2, and the blue curve is for MoS_2 on SiO_2, extracted from FETs. The mobility of MoS_2 is less dependent on the preparation process (blue curve; images from left to right: optical micrograph of MoS_2 made by CVD, AFM image of MoS_2 prepared by LPE, AFM image of MoS_2 prepared by MC) [36]

TMDC among other methods in the literature (Fig. 2.14h). Sodium incorporated in semiconductor devices is considered as contamination and can degrade device performance. Nevertheless, they are very affordable and provide high yield for large-area monolayers and will speed up the development of 2D semiconductors for the next-generation electronics and logics (Fig. 2.15).

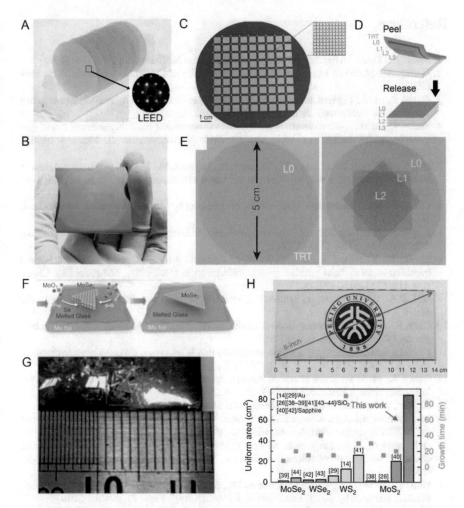

Fig. 2.15 (a) A tray of MoS_2 monolayer films grown on 2 inch c-sapphire. Low-energy electron diffraction pattern indicates the epitaxial nature of MoS_2 grown on sapphire [39]. (b) A 2 inch MoS_2 atomically thin film was transferred onto SiO_2/Si for complimentary metal-oxide-semiconductor applications [6]. (c) MOCVD and ALD processes for TMDC thin films can be independent of wafer sizes and provide excellent film uniformity, which is essential for massive device production [10]. (d, e) A variety of large-area TMDC films all grown via MOCVD can be mechanically transferred and stacked together using a thermal release tape (TRT) becoming a vdW heterostructure [40]. (f, g) $MoSe_2$ grows up to 5 mm on glass. During the growth at high temperatures, the melted glass releases sodium that promotes the radical growth rate [42]. (h) MoS_2 grown on 6 inch glass, which releases sodium during the growth, promoting the MoS_2 growth rate. Compared to previous methods, this work provides outstanding uniformity [43]

References

1. Smith, D.: Thin-film deposition: principles and practice. McGraw-Hill, New York (1995)
2. Ohring, M.: Materials science of thin films : deposition and structure. Academic, New York (2002)
3. Gupta, P., et al.: Layered transition metal dichalcogenides: promising near- lattice-matched substrates for GaN growth. Sci. Rep. **6**, 23708 (2016).
4. Zhang, C., et al.: Systematic study of electronic structure and band alignment of monolayer transition metal dichalcogenides in Van der Waals heterostructures. 2D Mater. **4**, 015026 (2016)
5. Das, S., Robinson, J.A., Dubey, M., Terrones, H., Terrones, M.: Beyond graphene: progress in novel two-dimensional materials and van der Waals solids. Annu. Rev. Mater. Res. **45**, 1–27 (2015)
6. Lin, Y.-C., et al.: Wafer-scale MoS_2 thin layers prepared by MoO_3 sulfurization. Nanoscale. **4**, 1–8 (2012)
7. Lee, Y.-H., et al.: Synthesis of large-area MoS_2 atomic layers with chemical vapor deposition. Adv. Mater. **24**, 2320–2325 (2012)
8. Li, H., Li, Y., Aljarb, A., Shi, Y., Li, L.-J.: Epitaxial growth of two-dimensional layered transition-metal dichalcogenides: growth mechanism, controllability, and scalability. Chem. Rev. **118**, 6134–6150 (2017)
9. Vilá, R.A., et al.: Bottom-up synthesis of vertically oriented two-dimensional materials. 2D Mater. **3**, 041003 (2016)
10. Kang, K., et al.: High-mobility three-atom-thick semiconducting films with wafer-scale homogeneity. Nature. **520**, 656–660 (2015)
11. Eichfeld, S.M., et al.: Highly scalable, atomically thin WSe_2 grown via metal-organic chemical vapor deposition. ACS Nano. **9**, 2080–2087 (2015)
12. Zhang, X., et al.: Influence of carbon in metalorganic chemical vapor deposition of few-layer WSe_2 thin films. J. Electron. Mater. **45**, 6273–6279 (2016)
13. de Heer, W.A., et al.: Epitaxial graphene. Solid State Commun. **143**, 92–100 (2007)
14. Van Bommel, A.J., Crombeen, J.E., Van Tooren, A.: LEED and Auger electron observations of the SiC(0001) surface. Surf. Sci. **48**, 463–472 (1975)
15. de Heer, W.A., et al.: Large area and structured epitaxial graphene produced by confinement controlled sublimation of silicon carbide. Proc. Natl. Acad. Sci. U. S. A. **108**, 16900–16905 (2011)
16. Forti, S., Starke, U.: Epitaxial graphene on SiC: from carrier density engineering to quasi-free standing graphene by atomic intercalation. J. Phys. D. Appl. Phys. **47**, 094013 (2014)
17. Lin, Y.-C., et al.: Direct synthesis of van der Waals solids. ACS Nano. **8**, 3715–3723 (2014)
18. Kroemer, H.: Heterostructure bipolar transistors and integrated circuits. Proc. IEEE. **70**, 13–25 (1982)
19. Van der Koma, A.: Waals epitaxy for highly lattice-mismatched systems. J. Cryst. Growth. **201–202**, 236–241 (1999)
20. Yang, W., et al.: Epitaxial growth of single-domain graphene on hexagonal boron nitride. Nat. Mater. **12**, 792–797 (2013)
21. Ago, H., et al.: Controlled van der Waals epitaxy of monolayer MoS_2 triangular domains on graphene. ACS Appl. Mater. Interfaces. **7**, 5265–5273 (2015)
22. Lin, Y.-C., et al.: Atomically thin heterostructures based on single-layer tungsten diselenide and graphene. Nano Lett. **14**, 6936–6941 (2014)
23. Lin, Y.-C., et al.: Atomically thin resonant tunnel diodes built from synthetic van der Waals heterostructures. Nat. Commun. **6**, 7311 (2015)
24. Gong, Y., et al.: Vertical and in-plane heterostructures from WS_2/MoS_2 monolayers. Nat. Mater. **13**, 1135–1142 (2014)
25. Li, M.-Y., et al.: Nanoelectronics. Epitaxial growth of a monolayer WSe_2-MoS_2 lateral p-n junction with an atomically sharp interface. Science. **349**, 524–528 (2015)

26. Shi, Y., et al.: van der Waals epitaxy of MoS$_2$ layers using graphene as growth templates. Nano Lett. **12**, 2784–2791 (2012)
27. Zhang, K., Lin, Y.-C., Robinson, J.A.: Semiconductors and semimetals, vol. 95, pp. 189–219. Elsevier, Amsterdam (2016)
28. Bradley, A.J., et al.: Probing the role of interlayer coupling and coulomb interactions on electronic structure in few-layer MoSe$_2$ nanostructures. Nano Lett. **15**, 2594–2599 (2015)
29. Tan, C., Zhang, H.: Epitaxial growth of hetero-nanostructures based on ultrathin two-dimensional nanosheets. J. Am. Chem. Soc. **137**, 12162–12174 (2015)
30. Ugeda, M.M., et al.: Giant bandgap renormalization and excitonic effects in a monolayer transition metal dichalcogenide semiconductor. Nat. Mater. **13**, 1091–1095 (2014)
31. Jung, Y., Shen, J., Sun, Y., Cha, J.J.: Chemically synthesized heterostructures of two-dimensional molybdenum/tungsten-based dichalcogenides with vertically aligned layers. ACS Nano. **8**, 9550–9557 (2014)
32. Kang, J., Liu, W., Sarkar, D., Jena, D., Banerjee, K.: Computational study of metal contacts to monolayer transition-metal dichalcogenide semiconductors. Phys. Rev. X. **4**, 031005 (2014)
33. Wang, Q.H., Kalantar-Zadeh, K., Kis, A., Coleman, J.N., Strano, M.S.: Electronics and optoelectronics of two-dimensional transition metal dichalcogenides. Nat. Nanotechnol. **7**, 699–712 (2012)
34. Yoon, Y., Ganapathi, K., Salahuddin, S.: How good can monolayer MoS$_2$ transistors be? Nano Lett. **11**, 3768–3773 (2011)
35. Radisavljevic, B., Radenovic, A., Brivio, J., Giacometti, V., Kis, A.: Single-layer MoS$_2$ transistors. Nat. Nanotechnol. **6**, 147–150 (2011)
36. Fiori, G., et al.: Electronics based on two-dimensional materials. Nat. Nanotechnol. **9**, 768–779 (2014)
37. Xie, L., et al.: Graphene-contacted ultrashort channel monolayer MoS$_2$ transistors. Adv. Mater. **29**, (2017). https://doi.org/10.1002/adma.201702522
38. Robinson, J.A., et al.: Epitaxial graphene transistors: enhancing performance via hydrogen intercalation. Nano Lett. **11**, 3875–3880 (2011)
39. Yu, H., et al.: Wafer-scale growth and transfer of highly-oriented monolayer MoS$_2$ continuous films. ACS Nano. **11**, 12001–12007 (2017)
40. Kang, K., et al.: Layer-by-layer assembly of two-dimensional materials into wafer-scale heterostructures. Nature. **550**, 229–233 (2017)
41. Li, S., et al.: Halide-assisted atmospheric pressure growth of large WSe$_2$ and WS$_2$ monolayer crystals. Appl. Mater. Today. **1**, 60–66 (2015)
42. Chen, J., et al.: Chemical vapor deposition of large-size monolayer MoSe$_2$ crystals on molten glass. J. Am. Chem. Soc. **139**, 1073–1076 (2017)
43. Yang, P., et al.: Batch production of 6-inch uniform monolayer molybdenum disulfide catalyzed by sodium in glass. Nat. Commun. **9**, 979 (2018)

Chapter 3
Properties of Atomically Thin WSe₂ Grown Via Metal-Organic Chemical Vapor Deposition

3.1 Impact of Growth Conditions and Substrates on Properties of WSe₂

3.1.1 Introduction

Two-dimensional tungsten diselenide (WSe$_2$) is of interest for the next-generation electronic and optoelectronic devices due to its bandgap of 1.65 eV and also its excellent transport properties. However, technologies based on 2D WSe$_2$ cannot be realized without a scalable synthesis process. The first part of this chapter focuses on the scalable synthesis for large-area, mono, and few-layer WSe$_2$ via metal organic chemical vapor deposition (MOCVD) using tungsten hexacarbonyl (W(CO)$_6$) and dimethylselenium ((CH$_3$)$_2$Se). In addition to the excellent scalability of production, this technique allows for the precise control of vapor-phase chemistry, which is not obtainable though the physical vapor reaction using powder precursors. Growth parameters such as temperature, pressure, Se to W ratio, and selection of the substrates for the growth play important roles on the resultant structure. With optimized conditions, domain size >8 μm is yielded.

Chemical vapor deposition of thick (>100 nm) but noncrystalline TMDC has been successful using a variety of metal-organics (W(CO)$_6$, Mo(CO)$_6$, etc.) [1–3] and metal-chlorides (MoCl$_5$, WCl$_5$, WOCl$_5$, VOCl$_5$) [1, 4–6] combined with a wide range of chalcogen precursors [1–6]. These early processes, while not refined to synthesize atomically thin layers, provide important insight into precursor chemistry ideal for the growth of monolayer TMDC and have led to a variety of reports on synthesis of monolayer MoS$_2$ [7–9], MoSe$_2$ [10–13], and WS$_2$ [14, 15]. Additionally, synthesis of WSe$_2$ has been reported via various techniques including pulsed laser deposition [16], amorphous solid-liquid crystalline solid [17], and powder vaporization (PV) [18–20]. These methods, while important for understanding the properties of monolayer TMDC, lack the control and reproducibility of the precursors needed

© Springer Nature Switzerland AG 2018
Y. -C. Lin, *Properties of Synthetic Two-Dimensional Materials and Heterostructures*, Springer Theses,
https://doi.org/10.1007/978-3-030-00332-6_3

for a truly scalable synthesis process. Thus, in order to advance technology, developing a scalable process that allows for more precise control of both the metal and chalcogen precursors is requisite.

This part presents one of the first MOCVD processes for synthetic WSe$_2$, including those on a wide variety of substrates including sapphire, graphene, and amorphous boron nitride (aBN), and also provides evidence how layer properties can be controlled by varying Se:W ratio. Characterizations using Raman spectroscopy, atomic force microscopy (AFM), and field emission scanning electron microscopy (FESEM) correlate domain size, layer thickness, and morphology of the synthetic WSe$_2$ atomic layers with the Se:W ratio. Growth conditions necessary to obtain large (5–8 μm) domains are discussed including the effect of temperature, pressure, and Se:W ratio. Conductive AFM (CAFM) and current-voltage (I_{ds}–V_{ds}) measurements on WSe$_2$/epitaxial graphene (EG) also provide evidence that the MOCVD process leads to electronic-grade heterostructures, suggesting a pristine interlayer gap is present between the WSe$_2$ and EG.

3.1.2   Experimental Methods

Material Synthesis

Tungsten selenide was synthesized using tungsten hexacarbonyl (Sigma Aldrich 99.99% purity) and dimethylselenium precursors (SAFC (99.99% purity) or STREM Chemical (99% purity)) in a vertical cold-wall induction-heated susceptor. The precursors were dispensed into the system via a bubbler manifold allowing for independent control over each precursor concentration. The carrier gas included H$_2$/ N$_2$ mixtures, with 100% H$_2$ being optimal. The samples were heated to 500 °C at 80 °C/min and annealed for 15 minutes to drive off any water vapor. Samples were then heated to growth temperature (600–900 °C) at 80 °C/min. Upon reaching growth temperature, the W(CO)$_6$ and DMSe were introduced into the reaction chamber. Growth took place at total pressures from 100 to 700 Torr and growth times were 30 minutes. The Se and W concentrations were varied by changing the H$_2$ carrier gas flow rate or bubbler temperature. Samples were cooled to room temperature. (Note: This process was developed by Dr. Sarah Eichfeld, Dr. Joshua Robinson, and Dr. Joan Redwing, with significant assistance provided by Ms. Lorraine Hossain, who is currently a graduate student at UCSD).

Epitaxial graphene is grown on diced SiC wafers via sublimation of silicon from 6H-SiC (0001) at 1700 °C for 15 min under 1 Torr Ar background pressure [19]; CVD graphene was prepared via a catalytic CVD method on 25-μm 99.999% pure Cu foils at 1050 °C, 1 Torr, and transferred onto SiO$_2$/Si via PMMA membrane [21]. Boron nitride was deposited on sapphire substrates via a pulse laser deposition (PLD) technique [22].

Materials Characterization

The as-grown samples are characterized using Raman spectroscopy, atomic force microscopy (AFM), and transmission electron microscopy (TEM). A WITec CRM200 Confocal Raman microscope with a 488 nm and 633 nm laser wavelength is utilized for structural characterization. A BRUKER Dimension 3100 with a scan rate of 0.75–1 Hz was utilized for the AFM measurements. The scanning electron microscopy was carried out on a Zeiss MERLIN FESEM. TEM cross-sectional samples were made by FEI Nova 200 dual-beam FIB/SEM with lift-out method. A carbon layer was deposited on the WSe$_2$ surface to avoid electron charging. In FIB, SiO$_2$ and Pt layers were deposited to protect the interested region during focused ion beam milling. A JEOL ARM200F transmission electron microscope operated at 200 kV with probe aberration corrector was used for high-resolution TEM (HRTEM) imaging and energy-dispersive X-ray spectroscopy (EDS) analysis.

Device Fabrication and Tunneling Current Measurements

The vertical diode was fabricated with electron beam lithography and lift-off of evaporated metal contacts. In the first step, the graphene contact is patterned and developed with electron beam (e-beam) lithography. Subsequently, metal contacts Ti/Au (10 nm/40 nm) are deposited with low-pressure electron beam evaporation (10^{-7} Torr) after an oxygen plasma treatment to reduce the contact resistance (45 s at 100 W, 50 sccm He, 150 sccm O$_2$ at 500 mTorr). Then a layer of 30 nm Al$_2$O$_3$ is deposited conformally over the entire substrate with atomic layer deposition (ALD), which serves as a protection layer for subsequent processing steps and a passivation layer. ALD deposited Al$_2$O$_3$ capping layer has been reported as an effective film to substantially block influence of ambient. In the second e-beam lithography step, a pattern of etch regions are defined, including an opening on the Ti/Au pads, and regions for the later WSe$_2$ contacts. The Al$_2$O$_3$ capping layer on these regions is first removed with hydrofluoric acid followed by oxygen plasma etching to remove the monolayer WSe$_2$ and few layers of graphene. This step prevents shorting through the underlying graphene layer after depositing the WSe$_2$ contacts. In the third e-beam step, the WSe$_2$ contact pads and thin lines are defined, and the Al$_2$O$_3$ layer on the WSe$_2$ triangular sheets is removed by hydrofluoric acid prior to the metal deposition. Then 50 nm thick palladium (Pd) layer is deposited by electron beam evaporation at 10^{-7} Torr. The high work function Pd contacts with WSe$_2$ have been reported to produce a smaller Schottky barrier and many orders higher current density compared to Ti/Au contacts.

3.1.3   Results and Discussion

The synthesis of WSe$_2$ was carried out via metal organic chemical vapor deposition (MOCVD) in a vertical, cold wall system using tungsten hexacarbonyl (W(CO)$_6$) and dimethylselenium (DMSe, (CH$_3$)$_2$Se) as the W and Se sources, respectively (Fig. 3.1a,b). The precursor purity has significant impact on the resultant film quality, where 99% pure (CH$_3$)$_2$Se exhibits much higher carbon contamination compared to 99.99%, regardless of H$_2$ concentration (Fig. 3.1c). While previous reports suggest adding small amounts of H$_2$ promotes WSe$_2$ growth [24], synthesis using

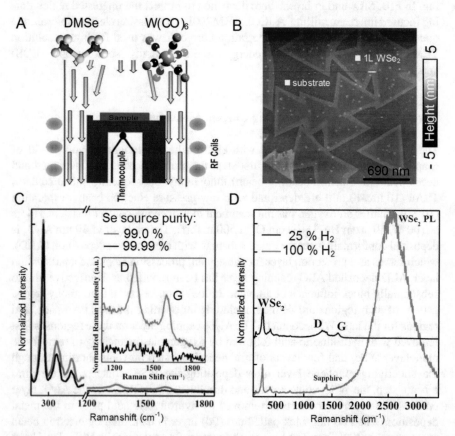

Fig. 3.1 (a) Schematic of MOCVD process allowing for precise precursor control in a vertical cold wall system for the investigation of the synthesis conditions. (b) AFM of WSe$_2$ on sapphire after growth showing monolayer WSe$_2$ was achieved. (c) The impact of the impurity in Se precursor on the WSe$_2$ monolayer under the same growth conditions. The red line indicating a Se source purity of 99.0% and the black curve indicating a purity of 99.99%: Raman spectra indicating that the Se precursor with higher impurity yielded carbon impurity incorporation in the WSe$_2$ layers. (d) The impact of H$_2$ on the growth of WSe$_2$ Raman spectra comparing 100% H$_2$ versus 1:3 H$_2$:N$_2$ as the carrier gas for synthesis of WSe$_2$. A H$_2$:N$_2$ mix for the carrier gas shows the carbon impurity as seen by D and G peaks. The PL is also quenched under the presence of carbon in the WSe$_2$ [23]

metal-organics requires the use of 100% hydrogen to minimize the carbon impurity incorporation from the $W(CO)_6$ and $(CH_3)_2Se$ precursors (Fig. 3.1d) [25].

The choice of substrate clearly has a significant impact on the morphology of atomically thin WSe₂ domains. This is apparent in Fig. 3.2, where AFM confirms that EG, CVD graphene, sapphire, and amorphous boron nitride substrates all yield distinct morphologies and thicknesses when grown under the same conditions. This suggests that there may be significant interaction between the WSe₂ and substrate during synthesis, even when the WSe₂ should have no dangling bonds out-of-plane when formed. Epitaxial and CVD graphene yield the highest nucleation density of monolayer WSe₂ domains, while amorphous boron nitride yields the lowest nucleation density with a strong preference for vertical (3D) growth of WSe₂ versus lateral (2D) growth. The presence of reactive defects and wrinkles in graphene is known to provide low-energy nucleation sites for the growth of MoS₂ [19]. This is also the case in this work for WSe₂, where graphene defects and surface contamination from the transfer process result in a high density of 3D-WSe₂ structures at the center of most 2D-WSe₂ domains. Growth on sapphire substrates yields the largest domains (5–8 μm) with additional layers growing from edge sites or defect sites on the monolayer. This suggests that the sticking coefficient for Se and W atoms on the surface of sapphire is greater than the other substrates, providing a means to achieve larger triangles through diffusion of source material across the substrate surface.

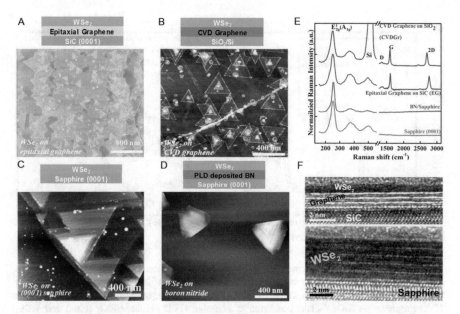

Fig. 3.2 (**a–d**) AFM scans showing differences in the WSe₂ morphology when grown on (**a**) epitaxial graphene, (**b**) CVD graphene, (**c**) sapphire, and (**d**) Amorphous boron nitride. (**e**) Raman spectra for synthetic WSe₂ on the various substrates showing similar quality. (**f**) Top: Cross-sectional TEM showing high-quality WSe₂ grown on epitaxial graphene. Bottom: Cross-sectional TEM of high-quality, multilayer WSe₂ on sapphire [23]

Finally, the presence of the E_{2g} and A_{1g} peaks of WSe₂ in Raman spectroscopy (see Fig. 3.2e) are observed, confirming the presence of WSe₂ [26]. Similar to previous reports [19, 27], the synthesis of the WSe₂ on graphene to form a vdW heterostructure does not appear to significantly degrade the underlying graphene based on the minimal "D" peak at 1360 cm⁻¹ in the Raman spectra.

Metal-organic chemical vapor deposition yields crystalline WSe₂ atomic layers with a tunable optical bandgap based on substrate choice. Cross-sectional transmission electron microscopy (TEM) of WSe₂ on EG and sapphire (Fig. 3.2f) confirms the presence of crystalline WSe₂ with pristine interfaces. In the case of growth on epitaxial graphene, TEM confirms the presence of three layers of epitaxial graphene and a single monolayer of WSe₂, with a clean interface and no observable defects. On the other hand, for the case of multilayer WSe₂ on sapphire, TEM reveals disorders at the WSe₂/sapphire interface suggesting a reaction during growth, which is similar to that found for WSe₂ films synthesized via selenizing tungsten oxide [28]. In the case of WSe₂ grown on CVD graphene, Raman spectroscopy provides evidence that, while no additional defects are found in the graphene after growth, there is a significant amount of strain introduced into the graphene following the deposition of WSe₂.

This was further investigated by examining shifts in the Raman 2D and G peaks in Fig. 3.3a. The data are vector decomposed (Fig. 3.3b) to correlate peak shifting to tensile and compressive strain ("eT" and "eC," respectively), Fermi velocity reduction (eFVR), and hole doping (eH), using methods by Ahn et al. Therefore, it

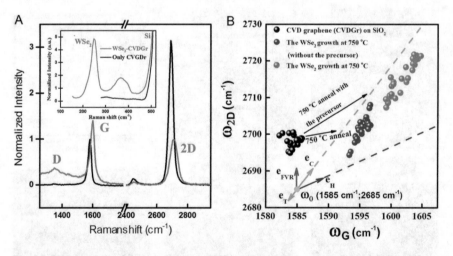

Fig. 3.3 The presence of strain in WSe₂ on CVD graphene. (**a**) Raman spectra of CVD graphene compared to WSe₂ on CVD graphene normalized to the SiO₂ at 520 cm⁻¹ showing significant G and 2D peak shifts to higher frequency. (**b**) Plot of Raman 2D frequency vs. G peak frequency for CVD graphene on SiO₂ (black) compared to annealed CVD graphene/SiO₂ (blue) and WSe₂ growth on CVD graphene/SiO₂ (red). The WSe₂ growth resulted in 0.4% compressive strain, while the same growth condition without W and Se sources introduced resulted in 0.2% compressive strain comparing to a freestanding graphene. This indicates that the WSe₂ deposited on CVD graphene contributes additional 0.2% strain in addition to the strain from the thermal effects. The strain in monolayer CVD graphene is calculated according to Ferralis et al. [23, 29]

is likely that the WSe$_2$ is also strained due to interlayer interactions, which ultimately reduces the bandgap by 30 meV [30, 31].

Growth conditions, including temperature and total pressure, dictate the overall domain size, domain geometry, and number of nucleation sites. Focusing on sapphire and EG, we find that the WSe$_2$ domain size increases with increased temperature and pressure. This is shown for growth on epitaxial graphene in Fig. 3.4a (the temperature is held at constant in the top row; and the pressure is held at constant in the bottom row). While the temperature is held constant at 750 °C, and Se:W ratio held at 100, the domain size increases from roughly 250 nm to 700 nm when the total pressure is increased from 500 to 700 Torr. Likewise, when the pressure is held constant at 650 Torr, and Se:W ratio held at 100, an increase in temperature of 100 °C (800 → 900 °C) yields a 200% increase in domain size (700 → 1500 nm). Synthesis at higher pressure also results in the formation of particulates on the sample surface, which were subsequently identified as W-rich WSe$_{2-x}$ nanoparticles via HRTEM (Inset in Fig. 3.4a). The presence of such particles indicates a lack of Se in the vapor phase during growth and therefore merited an investigation into the impact of Se:W ratio.

The Se:W ratio is critical in controlling defect formation in WSe$_2$. This is evident in Fig. 3.4b, where a surface plot of temperature and pressure versus Se:W ratio clearly demonstrates that domain size increases nontrivially as the Se:W ratio is increased to 800. Furthermore, as the Se:W ratio increases, there is a decrease in the density of W-rich WSe$_{2-x}$ particulates. This further supports the TEM analysis determining the particulates to be due to an imbalance in Se:W ratio and led to more detailed analysis of Se:W including "extreme" ratios. Figure 3.4c plots the domain size as a function of Se:W ratio. Extreme MOCVD ratios of up to 20,000 Se:W allows for a dramatic increase in domain size from 1 to 5 μm WSe$_2$ domains. We hypothesize that pushing the Se:W ratio to high values through a reduction in W(CO)$_6$ also leads to a decrease in the amount of nucleation sites and a reduced tendency to form Se vacancies which lead to secondary nucleation sites. Above a ratio of 20,000, however, the domain size begins to decrease again, suggesting that there is an ideal ratio for large domain growth.

Beyond temperature, pressure, and precursor ratios, the total flow through the system can also have a large impact. Figure 3.4d demonstrates the impact of total flow on the domain size and shape. Temperature, pressure, and Se:W ratio were held constant at optimized conditions (800 °C, 700 Torr, and 20,000 Se:W), while the total flow through the system was increased from 100 to 500 sccm. A total flow of 250 sccm yields 8 μm WSe$_2$ domains, while higher flow rates of 500 sccm result in a decrease in domain size and less defined WSe$_2$ edges. Increased total flow from 100 to 250 sccm increases the gas velocity in the system and leads to increased gas flux at the sample surface and higher lateral growth rates. However, increasing the total gas flow from 250 to 500 sccm leads to a decrease in domain size suggesting the gas velocity does not allow sufficient time for reaction of species at the substrate surface.

Since the family of vdW heterostructures is of increasing importance in the advancement of the field, synthesis of WSe$_2$ on graphene via MOCVD is also included in discussion. Comparing surface topography and conductivity acquired at V_{bias} = +0.8 V clearly indicates that an electrical barrier to transport exists in the area

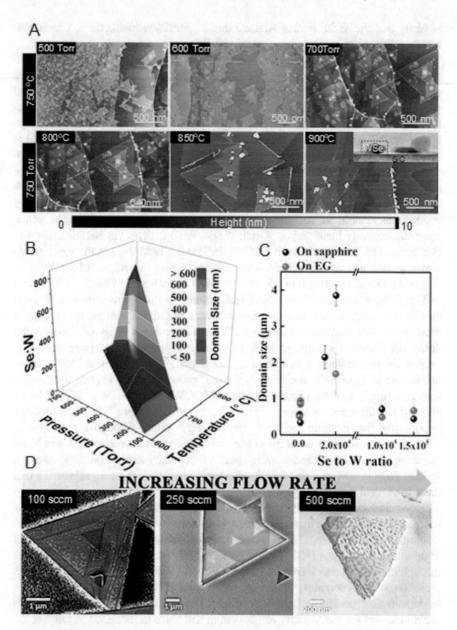

Fig. 3.4 (a) AFM of WSe₂ on EG showing increased domain size with increasing temperature and pressure. (b) Plot of temperature, pressure, and low (<1000) Se:W flux ratios as a function of domain size for sapphire substrates showing the impact of Se:W flux ratio on domain size. (c) Plot of extreme Se:W flux ratios as a function of domain size for both sapphire and epitaxial graphene indicating an optimum Se:W flux ratio around 2 × 10⁴. (d) FESEM showing the change in domain size as a function of total flow rate [23]

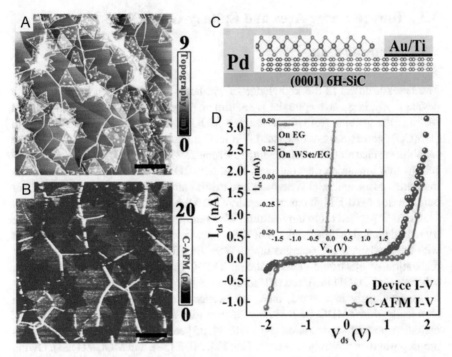

Fig. 3.5 (**a**) WSe₂-EG AFM correlated with (**b**) conductive-AFM mapping. (**c**) The schematic structure (side view) of 1 L WSe₂-EG diode using Au/Ti and Pt as source and drain contact, respectively. (**d**) WSe₂-EG diodes display a thermionic emission-like tunnel current turning on before 2 V, while WSe₂ grown on EG as a barrier reduced the current by 6 order (inset of **d**) [23]

of heterojunction. Current mapping reveals that both of 1 L WSe₂ and multilayer WSe₂ are resistive, while high conductivity is observed on the graphene, with graphene wrinkles (Bright stripes in Fig. 3.5b) exhibiting enhanced conduction in particular. Vertical diode structures (Fig. 3.5c,d) confirm the presence of a tunnel barrier created by the WSe₂ to vertical transport. The barrier is persistent under an increasing V_{bias} up to ±2 V prior to turn-on, while the area only having graphene clearly exhibits linear $I–V$ behavior (Inset, Fig. 3.5d). A resistance of ~10^{10} Ω prior to the turn-on suggests a significant electrical barrier at the WSe₂/graphene interface.

3.1.4   Conclusions for Sect. 3.1

This section was one of the first reports to achieve monolayer control of large domain WSe₂ via MOCVD, and this process allows for excellent control over the process conditions which is necessary to tune domain size, shape, and nucleation density. While MOCVD offers a highly scalable process with precise control over the gas phase chemistry, its precursor, DMSe unavoidably introduces carbon contamination into WSe₂.

3.2 Toward Large-Area and Epitaxy-Grade WSe$_2$

3.2.1 Introduction

The breakthroughs in the knowledge of synthesis and material process lead to a device-ready, large-area epitaxial WSe$_2$ film on a crystalline substrate. In the second section, the results from even cleaner MOCVD precursors for WSe$_2$ growth (Mo(CO)$_6$ and H$_2$Se) are discussed. We realize that when epitaxy is achieved, the sapphire surface reconstructs, leading to strong 2D/3D (i.e., TMDC/substrate) inter-actions that impact carrier transport. Even with 2D/3D coupling, transistors utiliz-ing transfer-free epitaxial WSe$_2$/sapphire exhibit ambipolar behavior with excellent on/off ratios (~10^7), high current density (1–10 μA μm^{-1}), and good FET mobility (~30 cm^2 V^{-1} s^{-1}) at room temperature. We examine the material interfaces and also try to understand the mechanisms of WSe$_2$ growth in order to correlate the transport with the fundamental properties of epi-WSe$_2$ thin films. This work establishes that realization of electronic-grade epitaxial TMDC must consider the impact of the substrate and 2D/3D interface as leading factors in electronic performance.

As previously mentioned, there are extensive efforts to synthesize large-area atomically thin TMDC by a variety of thin film deposition techniques [32, 33], including powder vaporization (PV) [18, 34, 35] selenization of W/WO$_x$ thin films pre-deposited on arbitrary substrates [36, 37], MBE [38], and MOCVD [23]. Owing to the ease of preparation and setup, PV using WO$_3$ and Se powders has been widely adapted to produce large WSe$_2$ domains with edge length between 1–100 μm on both crystalline and noncrystalline substrates [18, 27, 34]. Although the PV process provides the scientific community the fastest and most convenient route for obtain-ing materials for proof-of-concept demonstrations, its limited scalability will inevi-tably make itself obsolete for large-area commercial applications. On the other hand, MOCVD is an established manufacturing process for traditional compound semiconductors that enables the precise control of thermochemical reactions over an arbitrary size and time scales [39]. The capability to control the defect density and stoichiometry of synthetic TMDC by MOCVD is continually evolving and improving to the point where large-area films are readily available. Eichfeld et al. extensively studied MOCVD of WSe$_2$ using W(CO)$_6$ and DMSe and demonstrate that optimized growth windows utilize very high Se:W ratios for achieving large domains on sapphire and epitaxial graphene [23]. Similarly, Kim et al. synthesized large-area polycrystalline MoS$_2$ and WS$_2$, demonstrating uniform electrical and optical properties across 4-inch wafers [40]. In this study, by utilizing high-purity precursors and substrate surface engineering, we are able to achieve atomically thin epitaxial WSe$_2$ films with excellent chemical, structural, and elec-tronic uniformity.

Previously reported MOCVD process of WSe$_2$ mainly used carbon-contained MO precursors, such as DMSe or DESe as the Se source [23, 41]. In particular, the use of DMSe frequently leads to nano-size particles deposited on the surface of WSe$_2$ [23] or deposition of amorphous carbon layers at the interface between the substrate and WSe$_2$ [42]. Similarly, other MOCVD works for TMDC use alkali metal halide salts such as NaCl and KI to increase domain size and improve film morphology, which inevitably contaminate substrate and as-grown film [40, 43]. In order to suppress the carbon and other impurity contamination caused by the previous precursors, hydrogen selenide (H$_2$Se) was selected as the Se source for this work, while W(CO)$_6$ is the source for W. Instead of using the alkali metal halide salt, the reactor is filled with 100% H$_2$ during the growth to eliminate residual ambient molecules and also carbon contamination.

3.2.2   Experimental Methods

Materials Characterization

Atomic force microscopy (AFM) micrographs were taken with a Bruker Dimension at a scan rate of 0.5 Hz and 512 lines per image resolution. Peak Force KPFM mode using PFQNE AL probe on the same instrument was used to obtain the KPFM data. Lift height of 30 nm (or lower) and AC bias of 4 V were used during the surface potential measurement. The surface potentials of WSe$_2$ measured on the flat surface decreased monotonically with increasing layer number due to the enhanced screening of electronic trap states and dipole moments from the sapphire surface. The screening effect reaches a saturation on 3–4 L WSe$_2$ and is consistent with that observed on 1–4 L TMDs exfoliated on different surfaces. Scanning electron micrographs (SEMs) are taken in a LEO 1530 scanning electron microscope that uses a Schottky-type field-emission electron source and in-lens detector that receives the secondary electrons from the imaged sample. Raman and photoluminescence (PL) spectroscopy measurements (Horiba LabRam) were performed with 532 nm excitation wavelength, 100× objective lens.

Micro-size selected-area *low-energy electron diffraction* (LEED) is performed on WSe$_2$ films in an Elmitec III system. In the experiment, a collimating aperture can restrict the area of the electron beam to size between 1 and 7 μm in diameter, allowing us to locally probe the crystallinity across the surface.

Laser ablation inductively coupled plasma mass spectrometry (LA-ICPMS) with detection limit of 0.1 parts per million (ppm) was carried out by the Balazs NanoAnalysis lab on three different samples: sapphire substrate (used as background reference) and two WSe$_2$ films grown by DMSe and H$_2$Se. The impurities detected by LA-ICPMS are well below the X-ray photoemission spectroscopy (XPS) detection limit. XPS of all impurities as well as relevant core levels (W 4f, Se 3d, C 1 s, O 1 s, and Al 2p) were measured via a monochromated Al Kα source with takeoff angle of 45° and Omicron EA125 hemispherical analyzer with ±0.05 eV resolution and acceptance angle of 8° (26). The analyzer was calibrated with Au, Ag, and Cu foils according to "ASTM E2108-16 2016 *Standard Practice for Calibration of the Electron Binding Energy Scale of an X-Ray Photoelectron Spectrometer* (West Conshohocken, PA: ASTM)."

The STM/STS experiments were carried out with a cryogenic STM operated in ultrahigh vacuum at 5 K. Electrochemically etched tungsten tips cleaned in UHV by Ne ion bombardment and electron beam heating were used. STM images were recorded in constant current mode; bias voltages refer to the sample with respect to the STM tip. STS measurements of the differential tunneling conductance dI/dV were carried out with lock-in technique (modulation frequency 675 Hz at a peak-to-peak modulation of 10 mV) to probe the local density of electronic states. The STS spectra were recorded in a linearly varying tip height mode to improve the dynamic range in the bandgap region. All the spectra shown in the paper have been normalized to the constant tip height spectra.

To transfer WSe₂ onto a TEM grids or a fresh substrate, WSe₂/sapphire was spin-coated with 950 K PMMA using 2000 rpm for 30 s. After that, 16% HF solution was used to release PMMA/WSe₂ from sapphire. Once part of the substrate is merged in HF solution while the sample surface is above the solution surface, the solution will infiltrate into the interface between WSe₂ and sapphire substrate by capillary force to weaken the bond. The total releasing process takes around 1 min. At the end, the released specimen was positioned and captured on the target substrate and the supporting PMMA film was dissolved using acetone and isopropyl alcohol. All atomic resolution aberration-corrected HAADF STEM images were taken with a FEI Titan3 G2* with a resolution of 0.07 nm using 80 kV acceleration voltage.

Impurity of Precursor and Growth Steps of Epitaxial WSe₂

To understand the impact of precursor on film purity in MOCVD, WSe₂ films are deposited using two different chalcogenide precursors, dimethylselenium (DMSe, 99.99% purity) and hydrogen selenide (H₂Se, 99.9% purity), while keeping the metal precursor as tungsten hexacarbonyl (W(CO)₆). Utilization of DMSe leads to the deposition of carbon as confirmed by X-ray photoelectron spectroscopy (XPS) (Fig. 3.6a), which is likely due to an undesirable carbon layer that deposits on the sapphire simultaneous with growth of the WSe₂ film. Additionally, using ICPMS [5], we identify impurities including Fe, Na, Mg, Mo, Co, Zn, Cu, Au, Sb, As, Cr, Ga, Mn, Ni, Nb, Th, Sn, Y, Zr that are not detectable in XPS, indicating that the DMSe precursor exhibits high concentrations of impurities. After switching to a "carbon-free" chalcogenide precursor, H₂Se, the C-peak intensity is reduced by ~2–3x and the WSe₂ is stoichiometric. In addition, use of the H₂Se leads to significant reduction of impurity elements, with the complete elimination of Sb, As, Cr, Ga, Mn, Ni, Nb, Th, Sn, Y, and Zr (Fig. 3.6b) and reduction in Fe, Na, Mg, Mo, Co, Zn, Cu, and Au concentrations to $<10^{13}/cm^2$. Impurities may originate from the H₂Se, W(CO)₆, the growth chamber itself or post-growth sample handling.

There are three critical steps that we believe play an important role in achieving a high-quality epitaxial WSe₂ film including thermal treatment performed on as-received sapphire as the first step to create uniformly distributed step terraces; second, a "nucleation step" where the substrate is exposed to higher flow rates of W- and Se-precursors for a short period; lastly, post-growth annealing at 800 °C

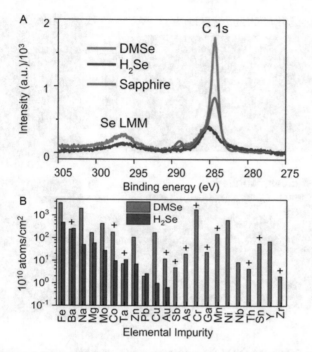

Fig. 3.6 (a) Binding energy of carbon measured in the XPS measurement found more carbon incorporated in WSe$_2$ grown with DMSe as Se source. (b) Elemental impurities detected in WSe$_2$ films using both H$_2$Se and DMSe measured by ICPMS. (plus) Marks the impurities that were also detected on the annealed c-plane sapphire used for growth [44]

under H$_2$Se flow to prevent Se vacancies from forming and also to reduce particles deposited on the surface. Verified by AFM and STM, this last step anneal reduces vacancies and surface nanoparticles. By following these steps, we are able to realize large-area epitaxial WSe$_2$ (Fig. 3.7) with highly uniform coverage and an excellent epitaxial relationship to the underlying sapphire substrate.

Electrolyte-Gating Electrical Measurement

The preparation of the polymer electrolyte is similar to previously published procedures by Xu et al. [45] with the exception that the electrolyte is prepared and deposited in an argon-filled glove box where the concentrations of H$_2$O and O$_2$ are maintained to <0.1 parts per million (ppm). Poly(ethylene oxide) (PEO) (molecular weight 95,000 g/mol, Polymer Standards Service) and CsClO$_4$ (99.9%, Sigma-Aldrich) are dissolved in anhydrous acetonitrile (Sigma-Aldrich) with an ether oxygen to Cs molar ratio of 76:1 to make a 1 wt% solution. The solid polymer electrolyte is deposited on the sample by drop-casting 25 µL onto the ~1 cm^2 sample, which is enough to cover a majority of the sample surface. After a 15 min wait time to allow the majority of the solvent to evaporate, the sample is annealed on a hotplate at

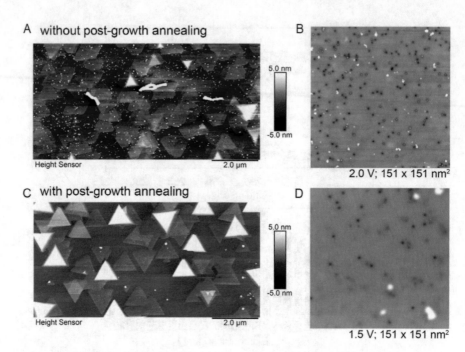

Fig. 3.7 (**a,c**) The topography of WSe₂ grains grown on annealed *c*-plane sapphire with (**a**) and without (**c**) post-growth annealing with H₂Se flow. (**b**) A STM view of WSe₂ monolayer grown on graphene without post-growth annealing. (**d**) An improved surface quality of WSe₂ grown on graphene after a 10-minute post-growth annealing was introduced. The applied voltage and scanned area are below their corresponding images [44]

80 °C for 3 min to drive off remaining solvent. The sample is then transferred from the glove box to the probe station through an Ar-filled load lock. The entire process of electrolyte preparation, deposition, transfer to the probe station, and measurement are completed under an inert gas environment with no sample exposure to ambient. Electrical measurements are performed on a Lake Shore cryogenic vacuum probe station (CRX-VF) under ~10⁻⁶ Torr at 300 K using a Keysight B1500A semiconductor parameter analyzer. Before applying the side gate bias (V_G), ions are homogeneously distributed in the PEO:CsClO₄.

The channel of the field effect transistor (FET) can be doped *n*- or *p*-type simply by applying voltages of opposite polarity to the side gate (Fig. 3.8a). When a positive V_G is applied, Cs⁺ ions are driven to the surface of the channel, which induces electrons in the WSe₂, turning it into an *n*-FET. Likewise, a *p*-FET can be realized by an opposite electrical field where ClO₄⁻ ions are driven to the channel and dope the channel inducing accumulation of holes. Due to the low mobility of ions in PEO at room temperature, a low sweep rate (8 mV s⁻¹) was used to provide sufficient time for the ions to respond to the applied electrical field (Fig. 3.8b). At the beginning of each transfer curve (I_D–V_G) measurement, S/D were grounded and V_G was held at 4 V for 5 min before initiating sweep to provide the ions sufficient time to

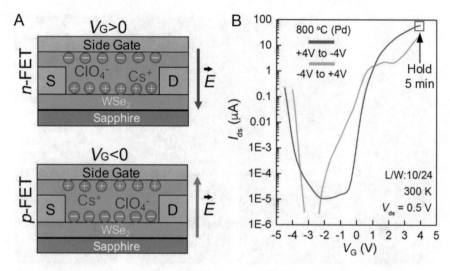

Fig. 3.8 (**a**) A schematic for the operation of *p*- and *n*-FETs using the solid electrolyte under $V_G > 0$ (Cs$^+$ ions form EDL on the channel) and $V_G < 0$ (ClO$_4^-$ ions form EDL on the channel), respectively. (**b**) Representative dual sweep during the transfer characteristic measurement on WSe$_2$ FET [44]

reach equilibrium at positive gate bias. All of the transfer curves in the following device study are then swept from positive V_G to negative V_G under a drain bias of 500 mV because this polarity (from 4 V to −4 V) provides a more ideal transfer curve where the on- and off-states can clearly be resolved.

3.2.3   Results and Discussion

Epitaxial Growth of WSe$_2$

Switching to H$_2$Se leads to dramatic improvement in WSe$_2$ purity compared to prior works, where carbon and many elemental impurities detected in inductively coupled plasma mass spectrometry (ICPMS) are eliminated. Furthermore, utilization of MOCVD provides a means to uniformly synthesize large-area monolayers across the substrate (Fig. 3.9a). The substrate and the subsequent surface preparation of the substrate is fundamentally important for achieving crystallographic alignment (epitaxy) and long-range order. While epitaxial growth of TMDCs is possible on graphene [27], gallium nitride [46], and sapphire [47, 48], *c*-plane sapphire [Al2O3 (0001)] is the substrate of choice in this study because it is a commercially viable [49] and highly chemically robust surface compatible with the harsh environments required for synthesis of TMDCs [47, 48]. The choice of sapphire, however, does not guarantee uniformity or epitaxy of the WSe$_2$. This is evident when growth is carried out on "as-received" *c*-plane sapphire, where we find optimized conditions

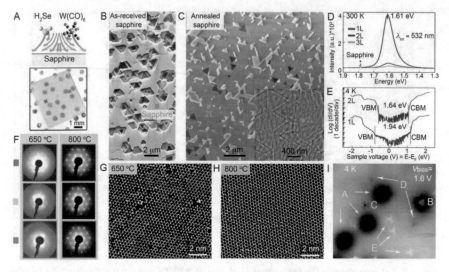

Fig. 3.9 (**a**) H₂Se and W(CO)₆ metal-organic precursors provide Se and W atoms in the growth of WSe₂. The bottom optical microscope image shows a uniform WSe₂ film grown on 1 cm² *c*-plane sapphire. (**b**) Growth performed on as-received sapphire at 800 °C resulted in isolated and multi-layered WSe₂ islands. (**c**) Scanning electron microscope image shows the same growth conditions performed on annealed sapphire achieved continuous 1 L WSe₂ film with the second layer aligned and starting to coalesce. Inset: An atomic force image of the clean morphology of the epitaxial WSe₂ film. (**d**) Evolution of the photoluminescence spectra obtained from the fully coalesced WSe₂ ranging from 1 L to 3 L. (**e**) Bandgap of 1 L and 2 L WSe₂ obtained in the STS measurement performed at 4 K. (**f**) Low-energy electron diffraction patterns obtained at three different spots on WSe₂ grown on annealed sapphire show a sharp hexagonal pattern on 800 °C WSe₂ but an unre-solved pattern on 650 °C WSe₂, indicating an improved epitaxial growth at higher growth tempera-ture. (**g,h**) The relationship between the quality of WSe₂ lattice in terms of defect density and the growth temperature was established in STEM measurement from (**g**) > 1 × 10¹⁴ cm⁻² at 650 °C to (**H**) ~1 × 10¹² cm⁻² at 800 °C. (**i**) Five most common types of defects observed in WSe₂ in STM measurement (**A**, Se-vacancy; **B**, W-vacancy substituted by a Se atom; **C**, double Se-vacancy, and **D/E**, impurity interstitials) [44]

often simply lead to multilayered, truncated domains (Fig. 3.9b) due to low surface energy and potential surface damage caused by mechanical polishing. However, annealing the sapphire substrate in air leads to the reconstruction of the surface resulting in terracing and regular atomic steps (0.2–0.4 nm tall and 50–300 nm wide) [48]. The reconstruction leads to a significant enhancement in surface energy, which enables uniform film coverage and layer-by-layer growth of the WSe₂ grown under the same growth conditions (Fig. 3.9c). The threefold symmetry of 2H-phase TMDCs and its long-range commensurability with the *c*-plane sapphire [47, 48] result in an epitaxial orientation of the WSe₂ grains that is either 0° or 60° compared to each other. Importantly, subsequent layer growth also follows this relationship. Furthermore, atomic force microscopy (AFM) verifies that the atomic steps of *c*-plane sapphire are readily translated to the WSe₂ topography (Fig. 3.9c, inset). The photoluminescence (PL) of the epitaxial WSe₂ clearly identifies a direct to indirect bandgap transition as the film thickness is increased from 1 L to 3 L, where the peak intensity is reduced and the energy positon red-shifted (Fig. 3.9d) [50]. Further investigation of the WSe$_2$ via scanning tunneling spectroscopy (STS) (Fig. 3.9e) confirms that the bandgaps of 1 L and 2 L WSe$_2$ grown on epitaxial graphene (EG)/SiC are 1.94 and 1.64 eV, which is consistent with the MBE-grown WSe$_2$ on EG [38].

While substrate surface engineering is requisite for epitaxy, it is not the only requirement. The growth temperature (T_G) is also a primary factor in achieving commensurability between WSe$_2$ and (0001) sapphire. This is clear when comparing the low-energy electron diffraction (LEED) patterns of WSe$_2$ films grown at 650 °C and 800 °C on 1 × 1 cm^2 annealed c-pane sapphire (Fig. 3.9f). The LEED patterns of films grown at 650 °C only exhibit a broad, blurred pattern, confirming the polycrystalline nature of the layers. On the other hand, samples grown at 800 °C exhibit clear hexagonal LEED patterns across the sapphire substrate, confirming the epitaxial nature of the layers despite the large lattice mismatch between the sapphire and WSe$_2$. This is possible via the commensurability of a 3 × 3 superlattice of WSe$_2$ and 2 × 2 superlattice of c-plane sapphire, which exhibits a relatively small lattice mismatch of approximately 4.0%. However, evident from the 650 °C growth, if T_G is not high enough, then the deposited atoms do not have enough surface energy for diffusion to crystallographically rearrange and align with the sapphire. In this research, temperatures >700 °C are necessary to achieve commensurability between TMDCs and sapphire substrates, thus we focus on 800 °C for epitaxial synthesis of WSe$_2$ on annealed sapphire substrates.

Growth temperature also controls defect density in a TMDC. Scanning transmission electron microscopy (STEM) provides direct evidence that the density of point defects within the lattice of WSe$_2$ is reduced by 100× (from $\sim10^{14}$ cm^{-2} to $\sim10^{12}$ cm^{-2}) when T_G is increased from 650 to 800 °C (Fig. 3.9g,h). Additionally, there is also a greater density of multiatom defects and defect complexes (dark regions) in WSe$_2$ grown at 650 °C compared to 800 °C. Scanning tunneling microscopy (STM) performed at 5 K on 1 L WSe$_2$ /EG identifies five types of defects with distinct features on the WSe$_2$ surface, labeled as A to E in Fig. 3.9i. Such defects adversely affect carrier transport in WSe$_2$ and thus need to be minimized. In order to reduce these defects, a 10-minute annealing in H$_2$Se flow was included after the growth step. During this post-growth annealing, H$_2$Se flow is maintained at 800 °C. The surface quality was significantly improved by post-growth annealing, as the numbers of particles and point defects on WSe$_2$ surface were reduced. Quantitatively, types A and B were reduced by $\sim10×$, and the high-quality hexagonal lattice becomes more evident, and in some areas, the type C defect was entirely eliminated. On the other hand, the density of defect types D and E remained the same level after the post-growth annealing. It is possible that the increased types D and E are due to excess Se deposited during growth cooldown or impurities incorporated during growth (i.e., iron from gas pipes and sodium from quartz tubes in the reactor). Nevertheless, the overall density of defects in an area of the sample that went through the post-annealing is around 7.5 × 10^{11} cm^{-2}, which is compatible to the densities from MBE-grown WSe$_2$ (2.8 × 10^{12} cm^{-2}) and mechanically cleaved WSe$_2$ crystals (1.2 × 10^{12} cm^{-2}) [38, 51].

Kinetics for WSe₂ Growth

There are several steps that need to be considered rather than a direct link from adatom mobilities to the epitaxy as well as improved domain quality. The growth temperature window in this study is 650–800 °C. Also, the total flow rate and the ratio of W to Se were constant. The growth temperature plays an important role in the kinetics of domain formation. It dictates the rates of absorption and absorption between gas and adatoms and the rates from adatoms to attachment (Fig. 3.10), which can be described by an Arrhenius equation ($\sim e^{-k_B T}$), where T is growth temperature, Q stands for the activation energy for a kinetic process, and k_B is the Boltzmann constant.

A high density of W adsorption at 600 °C led to high-density nucleation on sapphire and hence formed a continuous film with misoriented thicker layers, as shown in the AFM image of the 600 °C (Fig. 3.11). These small nuclei are triangular but may not necessarily register to the underlying sapphire. There are a few reasons for this: (1) at 600 °C the adatoms formed stable nuclei immediately because desorption is nearly negligible. Although the nuclei density is high, they are not necessarily all registry to the sapphire. (2) It is possible that adatoms formed irregular W and Se clusters on sapphire at lower temperature. Once they formed it is hard to reactivate them due to their strong binding energy (activation energy for W dimers is 5.44 eV and for Se dimers is 4.64 eV, while the activation energy for diffusion on a TMDC surface is only 1.34 eV and 0.23 eV for W and Se, respectively) [54]. These clusters would diminish the epitaxy if WSe₂ grow outward from their edges and require higher temperature to remove them. (3) According to first principle DFT calculation, the absolute value of the adsorption energy is typically higher than the diffusion energy barrier. This means that most adatoms diffuse to the WSe₂ domain edges and then attach, even the one that is unregistered to the sapphire, before they desorb.

However, once the substrate temperature increases, the desorption rate increases exponentially due to the Arrhenius nature of kinetic process and subsequently reduce adatom density and sticking coefficient. When the absorption and desorption is close to the equilibrium for adatoms at higher substrate temperature, attachment no longer dominates the reaction and the process of edge diffusion starts to be responsible to turn the domain into the thermodynamically favorable shape form of the initial randomness (Fig. 3.11a–c).

Additionally, reduced nucleation density and reduced surface imperfection at higher growth temperature would also improve the epitaxy of adatoms with the sapphire. Although the growth rate is slower at higher substrate temperature due to a competition of adsorption/desorption, as evident by our experimental observation and kinetic Monte Carlo simulations that incorporate possible kinetic processes, including nucleation, diffusion, and growth on the WSe₂ process from low to high temperature (Fig. 3.11d–f) [53], the film morphology can be significantly improved by tuning it from branching to compact by increasing the substrate temperature. Coincidently, the predicted morphology and its predicted growth time correspond-

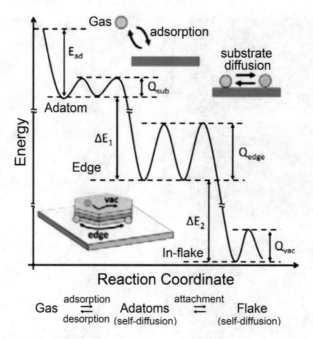

Fig. 3.10 The simulated reaction energy diagram of the WSe$_2$ growth process and its three-stage model. Each process is defined by its transition activation energy Q (i.e., Q_{sub} and Q_{edge}) [52]

ing to different growth temperature matches the experimental results well (Fig. 3.11g). Therefore, from the kinetic perspective, increasing temperature is necessary to improve the registry between the adatoms and substrate and subsequently improve the domain registry.

Domain Boundaries of Epitaxial WSe$_2$

Even with low defect density, epitaxial WSe$_2$, domain boundaries (DBs) can form as the films coalesce. Based on high-resolution STEM investigations, epitaxial WSe$_2$ DBs are predominately categorized as (1) the intersection and coalescence of domains that are oriented at 0° *and* atomically displaced in x or y or (2) antiphase domain boundaries that form when domains that are rotated 60° with respect to one another coalesce. Typically, when two aligned domains coalesce (0° DB), their joint interface predominantly remains hexagonal and uninterrupted (Fig. 3.12a,c). On the other hand, when domains are rotated 60°, a 4|4P boundary forms, referred to here as an antiphase domain boundary (APB), characterized by one Se atom coordinated with four W atoms (Fig. 3.12b,d) [55]. Such boundaries, in addition to containing a high density of atomic dislocations (Red squares in Fig. 3.12d), are predicted to be metallic [55].

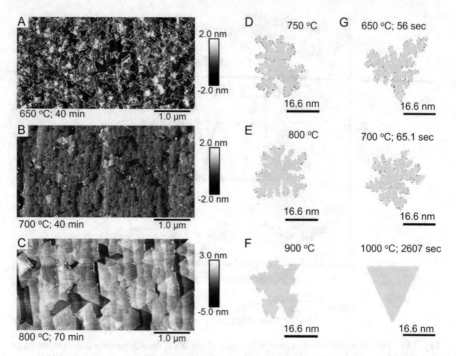

Fig. 3.11 (**a–c**) Show the images of WSe$_2$ grown for the temperature study, with other growth conditions fixed at constant. All of them have similar film coverage (~80%). (**d–f**) The monolayer WSe$_2$ domains formed at three temperatures predicted by the first principles Monte Carlo study, with the flux and W:Se ratio fixed at constant. (**g**) 3 WSe$_2$ domains with the similar size predicted by the Monte Carlo method and their predicted growth time. They were started with the same nucleus, metal flux, and W:Se ratio. The predicted time scale matches the experimental time scale for WSe$_2$ grown on sapphire at the three temperatures from low to high [44, 53]

WSe$_2$–Sapphire Interface

The synthesis of vdW materials on 3D substrates is not like traditional epitaxy, in part because vdW materials do not exhibit dangling bonds like that found in 3D. In order to achieve vdW epitaxial growth of 2D materials on traditional 3D substrates, Koma et al. passivated the surface dangling bonds with chalcogen atoms prior to layered TMDC growth (i.e., GaSe/Se-GaAs (111) and TX$_2$/S-GaAs) [56]. Epitaxy of WSe$_2$ on sapphire is accompanied by a "passivation layer" (noted as P) formation between the WSe$_2$ and sapphire substrate. A cross-sectional image of STEM (Fig. 3.13a) shows a vdW gap between the WSe$_2$ and sapphire surface. The sapphire surface also exhibits a structure different from that of the bare sapphire and now includes selenium (P in Fig. 3.13b), based on z-contrast intensity and the EDX mapping (Appendix Fig. A2). Density functional theory (DFT) modeling of various substrate surface terminations indicates that the passivating layer consists of selenium chains attached to the sapphire surface (Fig. 3.13c). The connection between the Se chain and the sapphire surface can be a direct Al-Se bonds or an Al-O-Se bridge via

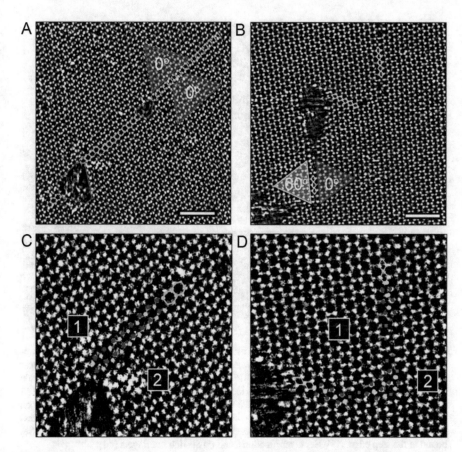

Fig. 3.12 (a) Z-contrast image of two domains aligned in the same direction that have coalesced and formed a seamless DB with the WSe₂ lattice remaining hexagonal (marked in light green). Voids are marked in purple. (b) The grains aligning against each other form 4|4P 60° DB (marked in orange). (c, d) Z-contrast images provide closer views on two major types of DBs. Different domains that form a DB are labeled by 1 and 2 (blue and yellow spheres represent W and Se atoms, respectively; dislocation is marked with red square) [44]

residual oxygen atoms on the surface (Appendix Fig. A3). The randomized orientation and length of the Se chains cause the blur of the STEM images. The simulated layer closely matches the interface identified by STEM imaging. Despite DFT modeling that predicts the presence of the passivation layer leads stronger WSe₂-substrate bonding, the electronic properties of the WSe₂ are not significantly altered, as the calculated WSe₂ density of states (DOS) shows no new energy states inside the bandgap caused by the passivating layer (Fig. 3.13c).

Substrate surface topography strongly influences the electronic properties of epitaxial WSe₂. Kelvin-probe force microscopy (KPFM) of an epitaxial WSe₂ film establishes that the topographic steps in the sapphire (Fig. 3.13d) induce localized modulation in WSe₂ surface potential (and hence Fermi level), creating a "striping"

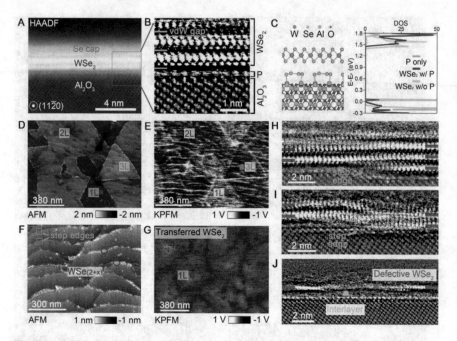

Fig. 3.13 (**a,b**) Indicates that the sapphire surface reconstructs to form a selenium-based passivating layer (P). DFT (**c**) indicates that the P consists of Al-Se that can lead to (**c**) energy states near the valance band edge (0 eV) of the WSe₂ DOS. Furthermore, AFM (**d**) and KPFM (**e**) provide evidence that the sapphire topography induces localized modulation of the WSe₂ Fermi level. AFM (**f**) of the original substrate indicates residual WSe₂ at the sapphire step edges, and KPFM (**g**) on the same film following transfer exhibits a 2–4× reduction in surface potential variation. Cross-sectional STEM confirms localized Fermi level modulations are likely the result of (**h**) WSe₂ bound to the step edge, (**i**) WSe₂ layer junctions, and (**j**) Se-rich interlayers [44]

effect in the KPFM map (Fig. 3.13e). The AFM of the original growth substrate following transfer (Fig. 3.13f) reveal residual stripes of Se-rich WSe₂ along the step edges, indicating that coupling between the WSe₂ and the step edge is much stronger than that found at the WSe₂/sapphire (0001) interface. The impact of the steps is further verified when the WSe₂ is transferred to a fresh SiO₂/Si (Fig. 3.13g), where the variability in potential is reduced by 2–4× and the "striping" has disappeared. Such modulation in the Fermi level and the presence of W/Se residue at step edges following film transfer provide direct evidence that the steps play a critical role in the growth and electronic transport of epitaxial WSe₂, making it essential to understand the physical source of this Fermi level variation.

Atomic steps in sapphire enable WSe₂ nucleation and induce structural variation. Evident from STEM (Fig. 3.13h,i), the first WSe₂ layer appears to nucleate at the sapphire atomic step edge and subsequently grows across the adjacent step edge and over the layer nucleating at that edge, similar to step-flow growth in traditional semiconductors [57]. Beyond providing a potential source for nucleation, the presence of the steps can lead to structural mixing of WSe$_2$ (Fig. 3.13i), which can also be accompanied by sporadic interlayers between sapphire and WSe$_2$ (Fig. 3.13j). Such structural bonding and mixing are correlated with the presence of a step edge, and are hypothesized to be the source of the Fermi level modulation, while the sporadic interlayers lead to circular bright spots in the surface potential map at terrace centers. Interestingly, the impact of the step edges is reduced with increasing layer thickness (Fig. 3.13e), indicating that each additional layer electronically screens the interface imperfections. Therefore, it is likely imperative to grow ≥ 2 L to effectively realize high-quality electronic transport and lateral device performance.

Electronic Transport of Epitaxial WSe$_2$

Electrolyte-gated bilayer WSe$_2$ FETs (Fig. 3.14a) demonstrate the impact of material properties on electronic performance [45, 58, 59]. FETs are evaluated for on/off ratio, subthreshold slope (SS), FET mobility (μ_{FET}), and threshold voltage (V_T). Prior to electrolyte deposition, the WSe$_2$ channel is highly resistive (Fig. 3.14b), suggesting the as-grown WSe$_2$ is p-doped such that the threshold voltage (V_G) is >0 V. Following electrolyte deposition the contacts are ohmic (Fig. 3.14b), indicating n-doping of the channel and thinning of the Schottky barrier at the WSe$_2$/metal interface [59]. Furthermore, transfer curves (I_{ds}–V_G) (Fig. 3.14c) indicate that palladium (Pd) contacts yield the highest μ_{FET}, best on/off ratio, and best SS for n-branch FETs. This is likely due to hybridization between Pd and WSe$_2$ surfaces that reduces the tunnel barrier at top of surface [60].

Growth temperature dramatically impacts epitaxial WSe$_2$ performance. This is evident when considering the transport of WSe$_2$ grown at 800 °C (^{800}WSe$_2$) and 650 °C (^{650}WSe$_2$) (Fig. 3.14c). Bilayer ^{800}WSe$_2$ exhibits colossal improvements in transport over ^{650}WSe$_2$, with ~1000× increase in on-current, 100–1000× higher on/off ratio (10^7), 100× higher μ_{FET} (~10 cm^2/Vs), and 2–3× lower SS for the n-branch (<200 mV/dec) in long channel FETs (Fig. 3.14c). In this case, the larger domains, reduced density of high-angle DBs, and dramatically reduced density of lattice point defects reduce impurity and phonon scattering. For ^{800}WSe$_2$, the p-branch cannot be resolved using the polymer electrolyte; however, based on the hole/electron μ_{FET} ratio for ^{650}WSe$_2$ (~10), we speculate that ^{800}WSe$_2$ could exhibit hole μ_{FET} as high as 100 cm^2/Vs at room temperature.

Substrate step edges act as doping and scattering centers in epitaxial TMDC. To determine the impact of atomic step edges, FETs are fabricated with channels parallel (FET$_\parallel$) and perpendicular (FET$_\perp$) to the step direction (Fig. 3.14d; dark stripes in the inset SEM image). Average μ_{FET} (and SS) for FET$_\parallel$ and FET$_\perp$ are 5.2 ± 0.7 cm^2/Vs (302 ± 50 mV/dec) and 3.8 ± 1.4 cm^2/Vs (322 ± 16 mV/dec), respectively. However, the V_T of FET$_\perp$ is shifted positive by >1 V, and the saturation current is nearly 2× lower, indicating the steps hole-dope the WSe$_2$ and scatter carriers at

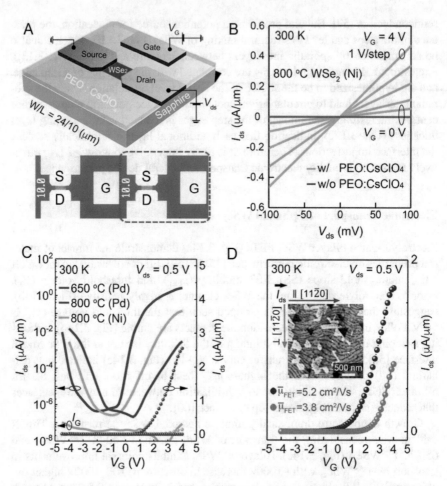

Fig. 3.14 Demonstration of WSe₂ FET: (**a**) Schematic details and optical image of the WSe₂ FETs. (**b**) I_d–V_d output characteristics indicate the electrolytic gate dopes the WSe₂ channel, improving contact resistance and shifting V_T. (**c**) Comparing transfer characteristics demonstrates superior performance of the 800 °C epitaxial WSe₂. (**d**) Transfer characteristics of WSe₂ channels parallel and perpendicular to substrate steps reveals the steps dope and scatter carriers [44]

higher rates than the (0001) Se-passivated sapphire plane. Furthermore, steps also lead to variation in the WSe₂ layer thicknesses, leading to modification in the band-gap thus requiring tunneling between WSe₂ layers to maintain electrical continuity. Importantly, however, the distribution of field-effect mobility, on/off ratio, and SS from devices across a 1 cm² WSe₂ film (Fig. 3.15) is highly uniform. Furthermore, comparing μ_{FET} versus current on/off ratio of all "large-area" synthetic WSe₂ films (Fig. 3.16) indicates that the ⁸⁰⁰WSe₂ with Pd contacts is comparable to the best single crystal bilayer WSe₂ domains reported, even though epitaxial WSe₂ exhibits smaller domains, domain boundaries, and many sapphire steps.

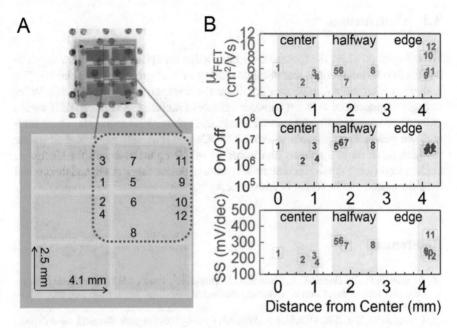

Fig. 3.15 (a) The layout shows the location of the 12 devices on a 1×1 cm^2 epi-bilayer WSe$_2$ film. (b) FET performance is uniform from the center to the edge when the channel is parallel to the substrate step edges

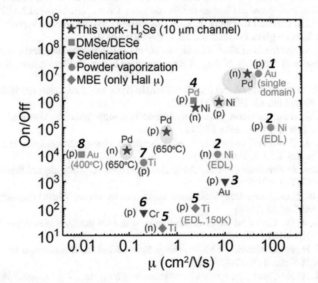

Fig. 3.16 Benchmarking state-of-the-art room-temperature device performance on synthetic WSe$_2$ compares the performance of epitaxial WSe$_2$ in this work. Only Hall mobilities are available for the WSe$_2$ grown on sapphire by MBE, while other mobilities are from room-temperature FET measurement [44]. [Reference in Fig. 3.16: (1) Nano Letter, 15, 709; (2) ACS Nano, 8, 923; (3) Nanoscale, 8, 2268; (4) 2D Materials, 3, 14,004; (5) Nano Letter, 17, 5595; (6) ACS Nano, 9, 4346; (7) Nanoscale, 7, 4193; (8) Journal of Electronic Materials, 45, 6280]

3.3   Conclusions

This chapter provides the foundational knowledge for epitaxy of WSe$_2$ on sapphire and the 2D/3D interactions that dominate transport in as-grown epitaxial layers. The realization that the substrate can dominate the transport of atomically thin WSe$_2$ strongly suggests that we must consider epitaxy of multilayer 2D materials if we are going to produce transfer-free, electronic grade, epitaxial 2D materials. These findings are generally applicable to other TMDCs and thus will guide and stimulate research interests in synthesis and transport of 2D epitaxial layers for electronic applications (more details on materials synthesis, device fabrication, and theoretical data/discussion can be found in Appendix A).

References

1. Boscher, N.D., Blackman, C.S., Carmalt, C.J., Parkin, I.P., Prieto, a.G.: Atmospheric pressure chemical vapour deposition of vanadium diselenide thin films. Appl. Surf. Sci. **253**, 6041–6046 (2007)
2. Chung, J.-W., Dai, Z.R., Ohuchi, F.S.: WS$_2$ thin films by metal organic chemical vapor deposition. J. Cryst. Growth. **186**, 137–150 (1998)
3. Hofmann, W.K.: Thin films of molybdenum and tungsten disulphides by metal organic chemical vapour deposition. J. Mater. Sci. **23**, 3981–3986 (1988)
4. Boscher, N.D., Carmalt, C.J., Palgrave, R.G., Gil-Tomas, J.J., Parkin, I.P.: Atmospheric pressure CVD of molybdenum Diselenide films on glass. Chem. Vap. Depos. **12**, 692–698 (2006)
5. Carmalt, C.J., Parkin, I.P., Peters, E.S.: Atmospheric pressure chemical vapour deposition of WS$_2$ thin films on glass. Polyhedron. **22**, 1499–1505 (2003)
6. Imanishi, N.: Synthesis of MoS$_2$ thin film by chemical vapor deposition method and discharge characteristics as a cathode of the Lithium secondary battery. J. Electrochem. Soc. **139**, 2082 (1992)
7. Lee, Y.-H., et al.: Synthesis and transfer of single-layer transition metal disulfides on diverse surfaces. Nano Lett. **13**, 1852–1857 (2013)
8. Schmidt, H., et al.: Transport properties of monolayer MoS$_2$ grown by chemical vapor deposition. Nano Lett. **14**, 1909–1913 (2014)
9. Yu, Y., et al.: Controlled scalable synthesis of uniform, high-quality monolayer and few-layer MoS$_2$ films. Sci. Rep. **3**, 1866 (2013)
10. Kong, D., et al.: Synthesis of MoS$_2$ and MoSe$_2$ films with vertically aligned layers. Nano Lett. **13**, 1341–1347 (2013)
11. Shim, G.W., et al.: Large-area single-layer MoSe$_2$ and its van der Waals heterostructures. ACS Nano. **8**, 6655–6662 (2014)
12. Wang, X., et al.: Chemical vapor deposition growth of crystalline monolayer MoSe$_2$. ACS Nano. **8**, 5125–5131 ((2014)
13. Chang, Y.-H., et al.: Monolayer MoSe$_2$ grown by chemical vapor deposition for fast photodetection. ACS Nano. **8**, 8582–8590 (2014)
14. Elías, A.L., et al.: Controlled synthesis and transfer of large-area WS$_2$ sheets: from single layer to few layers. ACS Nano. **7**, 5235–5242 (2013)
15. Tongay, S., et al.: Tuning interlayer coupling in large-area heterostructures with CVD-grown MoS$_2$ and WS$_2$ monolayers. Nano Lett. **14**, 3185–3190 (2014)

16. Grigoriev, S.N., Fominski, V.Y., Gnedovets, A.G., Romanov, R.I.: Experimental and numerical study of the chemical composition of WSex thin films obtained by pulsed laser deposition in vacuum and in a buffer gas atmosphere. Appl. Surf. Sci. **258**, 7000–7007 (2012)

17. Bozheyev, F., Friedrich, D., Nie, M., Rengachari, M., Ellmer, K.: Preparation of highly (001)-oriented photoactive tungsten diselenide (WSe$_2$) films by an amorphous solid-liquid-crystalline solid (aSLcS) rapid-crystallization process. Phys. Status Solidi. **211**, 2013–2019 (2014)

18. Huang, J.-K., et al.: Large-area synthesis of highly crystalline WSe$_2$ monolayers and device applications. ACS Nano. **8**, 923–930 (2014)

19. Lin, Y.-C., et al.: Direct synthesis of van der Waals solids. ACS Nano. **8**, 3715–3723 (2014)

20. Xu, K., et al.: Atomic-layer triangular WSe$_2$ sheets: synthesis and layer-dependent photoluminescence property. Nanotechnology. **24**, 465705 (2013)

21. Howsare, C.A., Weng, X., Bojan, V., Snyder, D., Robinson, J.a.: Substrate considerations for graphene synthesis on thin copper films. Nanotechnology. **23**, 135601 (2012)

22. Glavin, N.R., et al.: Amorphous boron nitride: a universal, ultrathin dielectric for 2D Nanoelectronics. Adv. Funct. Mater. **26**, 2640–2647 (2016)

23. Eichfeld, S.M., et al.: Highly scalable, atomically thin WSe$_2$ grown via metal-organic chemical vapor deposition. ACS Nano. **9**, 2080–2087 (2015)

24. Huang, J.-K., et al.: Large-area synthesis of highly crystalline WSe$_2$ monolayers and device applications. ACS Nano. **8**, 923–930 (2014)

25. Haigh, J., Burkhardt, G., Blake, K.: Thermal decomposition of tungsten hexacarbonyl in hydrogen, the production of thin tungsten-rich layers, and their modification by plasma treatment. J. Cryst. Growth. **155**, 266–271 (1995)

26. Terrones, H., et al.: New first order Raman-active modes in few layered transition metal dichalcogenides. Sci. Rep. **4**, 4215 (2014)

27. Lin, Y.-C., et al.: Atomically thin heterostructures based on single-layer tungsten diselenide and graphene. Nano Lett. **14**, 6936–6941 (2014)

28. Browning, P. et al.: Large-area synthesis of WSe$_2$ from WO$_3$ by selenium-oxygen ion exchange. 2D Mater. **2**, 1 (2014)

29. Ferralis, N., Maboudian, R., Carraro, C.: Evidence of structural strain in epitaxial graphene layers on 6H-SiC(0001). Phys. Rev. Lett. **101**, 156801 (2008)

30. Scalise, E., Houssa, M., Pourtois, G., Afanas'ev, V., Stesmans, A.: Strain-induced semiconductor to metal transition in the two-dimensional honeycomb structure of MoS$_2$. Nano Res. **5**, 43–48 (2012)

31. Castellanos-Gomez, A., et al.: Local strain engineering in atomically thin MoS$_2$. Nano Lett. **13**, 5361–5366 (2013)

32. Das, S., Robinson, J.A., Dubey, M., Terrones, H., Terrones, M.: Beyond graphene: progress in novel two-dimensional materials and van der Waals solids. Annu. Rev. Mater. Res. **45**, 1–27 (2015)

33. Shi, Y., Li, H., Li, L.-J.: Recent advances in controlled synthesis of two-dimensional transition metal dichalcogenides via vapour deposition techniques. Chem. Soc. Rev. **44**, 2744–2756 (2015)

34. Zhou, H., et al.: Large area growth and electrical properties of p-type WSe$_2$ atomic layers. Nano Lett. **15**, 709–713 (2015)

35. Huang, J., et al.: Large-area synthesis of monolayer WSe$_2$ on a SiO$_2$/Si substrate and its device applications. Nanoscale. **7**, 4193–4198 (2015)

36. Campbell, P.M., et al.: Field-effect transistors based on wafer-scale, highly uniform few-layer p-type WSe$_2$. Nanoscale. **8**, 2268–2276 (2016)

37. Chen, Y.-Z., et al.: Ultrafast and low temperature synthesis of highly crystalline and Patternable few-layers tungsten Diselenide by laser irradiation assisted Selenization process. ACS Nano. **9**, 4346–4353 (2015)

38. Zhang, Y., et al.: Electronic structure, surface doping, and optical response in epitaxial WSe$_2$ thin films. Nano Lett. **16**, 2485–2491 (2016)

39. Ohring, M.: Materials Science of Thin Films : Deposition and Structure. Academic, New York (2002)
40. Kang, K., et al.: High-mobility three-atom-thick semiconducting films with wafer-scale homogeneity. Nature. **520**, 656–660 (2015)
41. Park, K., et al.: Uniform, large-area self-limiting layer synthesis of tungsten diselenide. 2D Mater. **014004**, 3 (2016)
42. Zhang, X., et al.: Influence of carbon in metalorganic chemical vapor deposition of few-layer WSe$_2$ thin films. J. Electron. Mater. **45**, 6273–6279 (2016)
43. Kim, H., Ovchinnikov, D., Deiana, D., Unuchek, D., Kis, A.: Suppressing nucleation in metal-organic chemical vapor deposition of MoS$_2$ monolayers by alkali metal halides. Nano Lett. **17**, 5056–5063 (2017)
44. Lin, Y.-C., et al.: Realizing large-scale, electronic-grade two-dimensional semiconductors. ACS Nano. **12**, 965–975 (2018)
45. Xu, H., Fathipour, S., Kinder, E.W., Seabaugh, A.C., Fullerton-Shirey, S.K.: Reconfigurable ion gating of 2H-MoTe$_2$ field-effect transistors using poly(ethylene oxide)-CsClO$_4$ solid polymer electrolyte. ACS Nano. **9**, 4900–4910 (2015)
46. Ruzmetov, D., et al.: Vertical 2D/3D semiconductor Heterostructures based on epitaxial molybdenum Disulfide and gallium nitride. ACS Nano. **10**, 3580–3588 (2016)
47. Dumcenco, D., et al.: Large-area epitaxial monolayer MoS$_2$. ACS Nano. **9**, 4611–4620 (2015)
48. Chen, L., et al.: Step-edge-guided nucleation and growth of aligned WSe$_2$ on sapphire *via* a layer-over-layer growth mode. ACS Nano. **9**, 8368–8375 (2015)
49. Nakamura, S.: The roles of structural imperfections in InGaN-based blue light-emitting diodes and laser diodes. Science. **281**, 956–961 (1998)
50. Zhao, W., et al.: Lattice dynamics in mono- and few-layer sheets of WS$_2$ and WSe$_2$. Nanoscale. **5**, 9677–9683 (2013)
51. Addou, R., Wallace, R.M.: Surface analysis of WSe$_2$ crystals: spatial and electronic variability. ACS Appl. Mater. Interfaces. **8**, 26400–26406 (2016)
52. Yue, R., et al.: Nucleation and growth of WSe$_2$: enabling large grain transition metal dichalcogenides. 2D Mater. **4**, 045019 (2017)
53. Nie, Y., et al.: First principles kinetic Monte Carlo study on the growth patterns of WSe$_2$ monolayer. 2D Mater. **3**, 025029 (2016)
54. Nie, Y., et al.: A kinetic Monte Carlo simulation method of van der Waals epitaxy for atomistic nucleation-growth processes of transition metal dichalcogenides. Sci. Rep. **7**, 2977 (2017)
55. Zhou, W., et al.: Intrinsic structural defects in monolayer molybdenum disulfide. Nano Lett. **13**, 2615–2622 (2013)
56. Koma, A.: Van der Waals epitaxy for highly lattice-mismatched systems. J. Cryst. Growth. **201–202**, 236–241 (1999)
57. Xie, M.H., et al.: Anisotropic step-flow growth and island growth of GaN(0001) by molecular beam epitaxy. Phys. Rev. Lett. **82**, 2749–2752 (1999)
58. McDonnell, S., et al.: HfO$_2$ on MoS$_2$ by atomic layer deposition: adsorption mechanisms and thickness scalability. ACS Nano. **7**, 10354–10361 (2013)
59. Fathipour, S., Pandey, P., Fullerton-Shirey, S., Seabaugh, A.: Electric-double-layer doping of WSe$_2$ field-effect transistors using polyethylene-oxide cesium perchlorate. J. Appl. Phys. **120**, 234902 (2016)
60. Kang, J., Sarkar, D., Liu, W., Jena, D. & Banerjee, K.: A computational study of metal-contacts to beyond-graphene 2D semiconductor materials. International Electron Devices Meeting 17.4.1–17.4.4 (IEEE) (2012)

Chapter 4
Direct Synthesis of van der Waals Solids

4.1 Introduction

The stacking of two-dimensional layered materials such as semiconducting transition metal dichalcogenides (TMDCs), insulating hexagonal boron nitride (h-BN), and semi-metallic graphene has been theorized to produce tunable electronic and optoelectronic properties. In this chapter, we demonstrate the direct growth of MoS_2, WSe_2, and hBN on epitaxial graphene (EG) to form large-area van der Waal heterostructures. We reveal that the properties of the underlying graphene dictate properties of the heterostructures, where strain, wrinkling, and defects on the surface of graphene act as nucleation centers for lateral growth of the overlayer. Additionally, we demonstrate that the direct synthesis of TMDCs on EG exhibits atomically sharp interfaces. Finally, we demonstrate that direct growth of MoS_2 on EG can lead to a 10^3 improvement in photoresponse compared to MoS_2 alone.

Graphene is considered the foundation of exciting new science in two-dimensional layered materials [1]; but it is only the "tip of the iceberg." Novel device designs necessarily require additional high-quality film either as the barrier or active layer. Recently h-BN has attracted attention as a gate dielectric or substrate material for integration with graphene-based electronics as a gate dielectric or substrate material because its sp^2-hybridized bonding and weak interlayer vdW bonds result in a pristine interface [2]. This also leads to a decreased density of absorbed impurities that act as Columbic scatters when designing novel layered heterostructures [3–5]. Additionally, two-dimensional dichalcogenide-based materials are of significant interest for their finite bandgaps ranging from 3.5 eV for GaS [6] to <1 eV for $MoTe_2$ and WTe_2 [7]. More specifically, 2D TMDCs have gained momentum in recent years due to their applications in a variety of electronic and optoelectronic applications [3]. This is also complemented by the possibility to tune the energy bandgap of TMDCs from 0.8 to 2.1 eV through heterogeneous integration, thus producing entirely novel electronic and optoelectronic materials

© Springer Nature Switzerland AG 2018
Y. -C. Lin, *Properties of Synthetic Two-Dimensional Materials and Heterostructures*, Springer Theses,
https://doi.org/10.1007/978-3-030-00332-6_4

not yet synthesized [8]. These unique properties make TMDCs promising candidates for high-performance, low-cost energy materials for use in flexible electronics, photovoltaics [9], and energy storage [10, 11]. Development of electronically tunable vdW solids must start with high-quality substrate materials. Graphene and graphite are excellent templates for the growth of bilayer AB-stacked graphene [12], topological insulators [13], and other 2D materials such as h-BN and MoS_2 [14–18].

To date, progress in the development of vdW heterostructures has led to a variety of new phenomena [18]. However, these vdW structures are primarily fabricated via mechanical exfoliation using polymer membranes and micromanipulators to stack the individual 2D crystals [4]. The process of mechanical exfoliation, while often useful for demonstration purposes, can lead to interface contamination [19]. These defects and adsorbates buried at the interface of the 2D crystals undoubtedly diminish the quality and performance of devices. Therefore, the development of a growth technique to assemble these systems during synthesis is requisite for large-area, high-quality vdW solids. Several groups have demonstrated direct growth of bilayer vdW solids (two dissimilar layers) as the building block for further hetero-integration. Liu et al. [14] used a two-step ex-situ process to grow chemical vapor deposition (CVD) graphene followed by synthesis of h-BN on the CVD graphene. Similarly, Shi et al. [15] utilize CVD graphene as the growth template to grow MoS_2 by flowing $(NH_4)_2MoS_4$ precursor. While these are significant advances in the pursuit of vdW solids, the use of CVD graphene on Cu requires sophisticated methods of Cu-etching and transferring to avoid pinholes, tears, or surface contamination in the heterogeneous structures. This may limit the applicability of heterostructures based on CVD graphene simply due to variation in electronic properties due to polymeric contamination, mechanical strains, and substrate/vdW solid interface imperfections.

In this work, we utilize EG on 6H-SiC as the growth template for direct growth of MoS_2, WSe_2, and hBN. Quasi-freestanding EG (QFEG), a hydrogen-treated EG [20], is also utilized as a growing template and compared to EG to understand the impact of the EG buffer layer. Epitaxial graphene is utilized because it provides several technological advantages: (1) graphene is already on an insulating substrate, requiring no transfer processes, (2) the interface between graphene and SiC is pristine and tailorable, (3) the surface is free of polymeric and other contaminants found in transferred CVD graphene [21], and (4) EG is typically quite robust under standard device fabrication processes [22]. However, there are also challenges to utilization of epitaxial graphene (uniform thickness over large areas, steps in the SiC surface) that must also be considered. Here, we utilize atomic force microscopy (AFM), Raman spectroscopy, transmission electron microscopy (TEM), X-ray photoelectron spectroscopy (XPS), and photocurrent measurements to elucidate the impact of EG properties on the resulting MoS_2/EG, h-BN/EG, and WSe_2/EG heterostructures.

4.2   Experimental Methods

4.2.1   Materials Synthesis

Epitaxial graphene is grown on diced SiC wafers via sublimation of silicon from 6H-SiC (0001) at 1700 °C for 15 min under 1 Torr argon (Ar) background pressure. Quasi-freestanding graphene is prepared by exposing EG to 600 Torr hydrogen (H_2) at 1050 °C for 120 min to intercalate hydrogen at the graphene/SiC (0001) interface [22]. Growth of MoS_2 layers was accomplished using MoO_3 powders (0.1 g) placed in a ceramic crucible located in the center of a 2″ tube furnace. Sulfur powders are placed in the second ceramic crucible up stream and held at 130 °C during the reaction. The EG/SiC for growing MoS_2 were put at the downstream side. The MoO_3 and S vapors were transported to the EG/SiC substrates by pure Ar flowing gas (Ar = 50 sccm, chamber pressure = 5 Torr). The heating zone was heated to 670 °C at a ramping rate of 15 °C/min. Growth of WSe_2 layers on EG was accomplished by first thermally evaporating 5 nm WO_3 on EG/SiC. The WO_3/ QFEG/SiC was subsequently exposed to selenium vapor by heating pure selenium metal to 500 °C upstream in the tube furnace. This process converts the WO_3 to WSe_2, as discussed elsewhere [23]. Growth of hBN layers on QFEG was accomplished in a 75 mm diameter horizontal tube furnace via thermal CVD method utilizing ammonia borane (NH_3BH_3) precursor. Solid ammonia borane powder is sublimed at 135 °C and transported into the tube furnace by H_2/Ar carrier gas (5% of total flow rate). Growth occurs at 1075 °C and 250 mTorr in 5 min. After the growth, the ammonia borane carrier gas is turned off, and the furnace is allowed to cool down to room temperature slowly in a 250 mTorr Ar/H_2 environment.

4.2.2   Fabrication and Measurement of MoS₂ Photosensors

Two terminal photosensor devices were fabricated using standard ultraviolent photolithography. The sensors are with various source-drain spacing, which ranges from 1 to 15 μm. Titanium/gold (30/100 nm) ohmic contacts were deposited in a similar fashion to our graphene devices, which utilizes an oxygen plasma pretreatment [24]. Photocurrent measurements were carried out at room temperature in ambient and were coupled to a Renishaw micro Raman spectroscopy with a 488 nm laser. The electrical conduction data was collected with a power source and a Keithley 2400 semiconductor analyzer.

4.2.3   Materials Characterization

The as-grown heterostructures are characterized using Raman spectroscopy, atomic force microscopy (AFM), X-ray photoelectron spectroscopy (XPS), and transmission electron microscopy (TEM). A WITec CRM200 Confocal Raman microscope with a 488 nm laser wavelength is utilized for structural characterization. For as-grown epitaxial graphene, the SiC background signal is subtracted using a direct subtraction of the SiC substrate from the spectra [25]. A BRUKER Dimension with a scan rate of 0.5 Hz was utilized for the AFM measurements. A Kratos Axis Ultra XPS system utilizing an Al k-alpha source with energy of 1486.7 eV was used for XPS analysis. The TEM cross-sectional samples were made via utilizing a NanoLab dual-beam FIB/SEM system. Protective layers of SiO_2 and Pt were deposited to protect the interesting region during focused ion beam milling. TEM imaging was performed using a JEOL 2100F operated at 200 kV.

4.3   Results and Discussion

The utilization of epitaxial graphene provides a near perfect template for synthesis of VdW solids due to its lack of dangling bonds, chemical inertness, and ability to remain intact under high stress. Additionally, many of the 2D layered materials are isostructural, belonging to the symmetry group $P6_3/mmc$ [26]. However, based on the lattice parameters of graphene, hBN, MoS_2, and WSe_2 (2.461 Å, 2.50 Å, 3.16 Å, and 3.28 Å respectively), one may not expect epitaxial growth to proceed for transition metal dichalcogenides (TMDC) on graphene because a significant lattice mismatch exists. While this mismatch can be highly detrimental in 3D semiconductor epitaxy, there is a significant relaxation in the required lattice matching when growth proceeds via vdW interaction [27]. Density functional theory (DFT) modeling clearly indicates poor commensurability between strain-free graphene and MoS_2 (Fig. 4.1a,c), where the red dashed box is a guide to the eye illustrating the closest lattice match occurs at 3 and 4 unit cells (3–4) for MoS_2 and graphene, respectively. Figure 4.1c further illustrates the poor commensurability of the MoS_2/graphene (and other TMDC/graphene combinations) heterostructures at the (3–4) matchup. However, the weak vdW interaction provides a route for "epitaxy" to occur in these structures. Ongoing density functional theory (DFT) simulations, with vdW corrections as implemented in ONETEP [25], have already shown homogeneous-symmetrical reconstruction of the layered structures by inducing large strain on the layer of graphene at the local density approximation level of exchange and correlation functional [26]. Additionally, DFT indicates that the residual strain in epitaxial graphene (typical for graphene on SiC) [28] could impact epitaxy of the TMDC by providing a route for improved commensurability (Fig. 4.1b,d,e). Synthesis of high-quality epitaxial graphene (EG) was prepared by using an in situ hydrogen etch, followed by Si sublimation from Si face of semi-insulating 6H-SiC

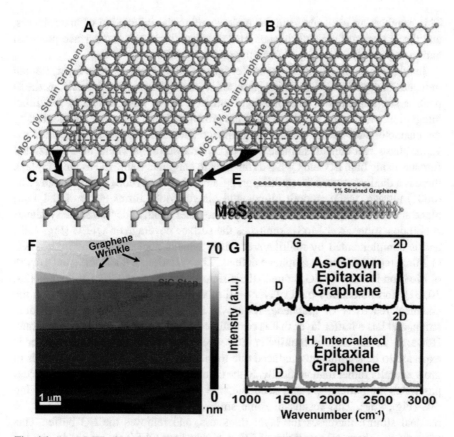

Fig. 4.1 (**a–e**) DFT of MoS$_2$/unstrained graphene and MoS$_2$/1% strained graphene predicts that residual strain in EG may enhance the structural symmetry in TMDC/graphene heterostructures. (**f**) AFM image of as-grown epitaxial graphene demonstrating a smooth surface along with 5–10 nm steps in the SiC substrate. Wrinkles in graphene grown on SiC (0001) appear due to the CTE mismatch between graphene and SiC. (**g**) Raman spectra of as-grown epitaxial graphene and H$_2$- treated (hydrogenated) epitaxial graphene. Following hydrogenation, the D-peak in EG is nearly eliminated by buffer layer passivation [29]

(II–VI, Inc.) at 1700 °C, 100 Torr. Some EG samples were further exposed to molecular hydrogen (hydrogenation) at 1050 °C, 600 Torr for 60 min to passivate the graphene/SiC (0001) buffer layer – referred as quasi-freestanding epitaxial graphene (QFEG). Hydrogenation typically results in the partial relaxation of residual strain in epitaxial graphene, which is evident in the red-shifted 2D Raman peak of the QFEG (Fig. 4.1d). Raman and transmission electron microscopy confirm a uniform, continuous top layer of epitaxial graphene (EG) extending over the SiC (1 $\bar{1}$ 0n) step edge and onto SiC (0001) terraces, with few-layer EG existing on the step edge after the growth. The QFEG exhibits atomically smooth surfaces on the SiC terrace, as well as atomic-scale wrinkles that result from the coefficient of thermal expansion (CTE) mismatch that exists between EG and SiC (Fig. 4.1f).

This provides an ideal platform for understanding the interaction of heterolayers, and the impact surface defects such as wrinkling, thickness, and surface potential variation have on the ability to form pristine vdW solids.

In addition to transforming the buffer into an additional graphene layer, Raman indicates the quality of QFEG is significantly improved compared to EG, as the D peak is suppressed. EG and QFEG are utilized as the base templates for direct integration with MoS_2. The nucleation and growth of MoS_2 is strongly influenced by the characteristics of the underlying graphene. Heterolayers were synthesized via vapor-phase reaction of sublimed MoO_{3-x} and sulfur powders in a horizontal furnace, using both EG and QFEG as the growth template. Using EG as the template, there is a distinct pattern for multilayer, pyramidal-shaped MoS_2 to form on the SiC (0001) terrace, while smooth mono- and bilayer MoS_2 forms at the SiC(1$\bar{1}$0n) plane and extends outward (Fig. 4.2a). On the other hand, under the same synthesis conditions, monolayer MoS_2 dominates the surface coverage on QFEG (Fig. 4.2b) and is complemented by multilayered MoS_2 islands nucleating on wrinkles, SiC (1$\bar{1}$0n) step edge, and at graphene defects. The variation in nucleation and growth of MoS_2 on EG and QFEG is due to the difference in graphene strain, thickness [25, 30, 31], and potentially buffer layer – all of which have significant impact on the chemical reactivity of graphene. As-grown EG is thinner, exhibits 1% residual strain, and has a buffer layer that is partially covalently bound to the SiC substrate. However, in the case of multilayer EG, where the impact of the buffer layer is expected to be significantly reduced due to increased EG thickness, MoS_2 tends to grow laterally rather than vertically. In the case of QFEG, the hydrogenation process passivates the buffer layer and decouples the graphene layers from the underlying SiC (Fig. 4.3a). This results in some strain relief of the graphene (~ 0.3–0.5% residual strain), increases the layer thickness, and removes the EG buffer. This suggests the chemical reactivity of EG is higher than QFEG due to strain and the presence of a buffer layer in EG, which results in a high density of TMDC nucleation sites on EG.

The MoS_2/EG (or QFEG) heterogeneous structures exhibit high-quality structural, chemical, and interfacial properties (Fig. 4.2c,d). Upon synthesis, the samples display MoS_2 peaks at 386 cm^{-1} and 408 cm^{-1} (E_{2g}^1 peak and A_{1g} peak, respectively) [32] and EG peaks at 1590 cm^{-1} (G peak) and 2740 cm^{-1} (2D peak) in the Raman spectra (Fig. 4.2c and its inset). The intense E_{2g}^1 and A_{1g} vibration modes of the MoS_2 Raman spectra indicate the MoS_2 is of high quality [32]. The E_{2g}^1 and A_{1g} peak spacing is also an indicator of layer thickness in TMDCs [32, 33] and is found to be 20.0 cm^{-1} for monolayer MoS_2 on EG. This value is 2 cm^{-1} larger than that in exfoliated monolayer MoS_2 and could be due to the stiffening of the A_{1g} mode with in situ synthesis and CTE mismatch [33]. Photoluminescence (PL) from the MoS_2/EG(QFEG) heterostructures also provides evidence that the films are of high crystalline quality and can range from mono- to few layer (Fig. 4.2d) under current growth conditions. Importantly, we note that Raman spectroscopy also confirms the direct synthesis of MoS_2 on EG and QFEG does not impact the quality of the underlying graphene (Fig. 4.2c shows no "D" peak). The element analysis on MoS_2/EG and MoS_2/QFEG via XPS also confirms a proper stoichiometric ratio

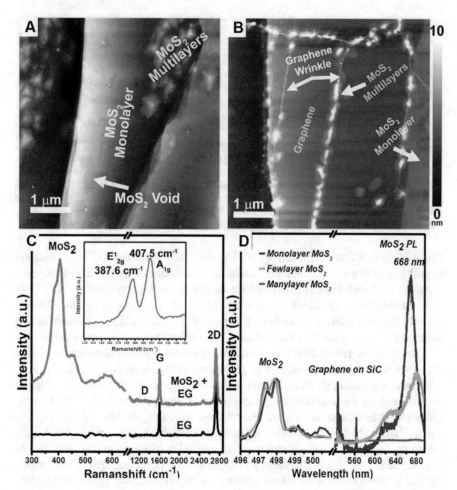

Fig. 4.2 Atomic layers of MoS$_2$ grown on (**a**) EG and (**b**) QFEG demonstrating a clear difference in nucleation and growth for two different graphene templates, which is attributed to the presence of enhanced residual strain and a buffer layer in EG compared to QFEG. In the case of QFEG, MoS$_2$ primarily nucleates on wrinkles and SiC step edges. (**c**) Raman spectra of MoS$_2$/EG and EG before the direct growth. Raman indicates that synthesis of MoS$_2$ does not induce additional defects in the EG or QFEG; Inset in (**c**) shows the E$_{2g}$ and A$_{1g}$ of the MoS$_2$ Raman features. (**d**) Photoluminescence from multilayer MoS$_2$ to monolayered MoS$_2$ after the growth demonstrating high-quality mono-, bi-, and multilayer MoS$_2$ is possible on epitaxial graphene [29]

between Mo and S of two, with no carbon bonding (Fig. 4.3a). The Mo 3d exhibits two peaks at 229.3 eV and 232.5 eV (Fig. 4.3b), which is attributed to the doublet 3d$_{5/2}$ and 3d$_{3/2}$, respectively. The peaks corresponding to sulfur 2p$_{1/2}$ and 2p$_{3/2}$ orbitals in bonded sulfide are observed at 164 eV and 162.5 eV (Fig. 4.3c). It should be noted that the resulted shapes of Mo and S in XPS are very similar, and thus only MoS$_2$/QFEG heterostructure case is presented. Following nucleation, there are distinct growth morphology differences between EG and QFEG. As noted in

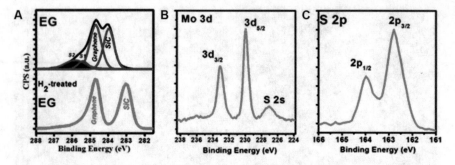

Fig. 4.3 X-ray photoemission spectroscopy demonstrates (**a**) that hydrogenation of epitaxial graphene on SiC (0001) eliminates the buffer layer (identified as S1 and S2) at the interface of graphene and SiC (0001). High-resolution XPS of the Mo 3d (**b**) and S 2p (**c**) peaks MoS$_2$, indicating no Mo-C or C–S bonding following MoS$_2$ synthesis on EG and QFEG templates [29]

Fig. 4.2a, there exists a high density of MoS$_2$ islands on the terrace of EG, which are hypothesized to grow in a manner similar to that described by Stranski-Krastanov (layer-plus-island) growth mode, which describes layer-plus-island growth [34]. Here, the MoS$_2$ films form via a combination of layer-by-layer growth and 3D island growth, which is similar to synthesis of graphene on h-BN [36]. The source of this islanding phenomenon may be related to the vertical propagation of a defect that initiates in the graphene layer (vacancy, strain-induced high chemical reactivity site, etc. – we later discuss the impact of graphene quality on MoS$_2$), which forms an additional nucleation site for adlayers to form and grow laterally outward from the central peak of the island. This phenomenon is also present in the QFEG case, although fewer isolated islands are identified on these samples where pristine QFEG exists. Rather, the QFEG wrinkles are more reactive due to the curved sp^2 π-bonds, which again induce localized strain and modification of the chemical reactivity of the graphene films [37]. The same phenomenon also occurs on QFEG step edges, similar to the synthesis of MoS$_2$ ribbons on highly oriented pyrolytic graphite surface (HOPG) [38]. Kelvin potential force microscopy measurements confirm the step edge, and the region adjacent to the step edge, where thicker graphene exists, exhibits a lower surface potential than that of graphene on the SiC terrace, confirming graphene on the SiC step edge is more reactive [29].

The impact of MoS$_2$ synthesis on the quality of QFEG is correlated with the measured layer thickness of the MoS$_2$ overlayer. Evident in Fig. 4.4, there are regions of high I_D/I_G ratios in the Raman spectra (Fig. 4.4a,b), which closely match the presence of wrinkles in the graphene film, as well as areas of "thick" MoS$_2$ (Fig. 4.4c–e). While wrinkles are present in as-grown QFEG, the I_D/I_G is <0.05 indicating that high-quality QFEG exists prior to MoS$_2$ synthesis. It is not until after the synthesis of MoS$_2$ that significant degradation occurs at regions of the QFEG at wrinkle locations, which also correlates with broadening of the 2D peak (Fig. 4.4b). Correlating I_D/I_G (Fig. 4.4a) with the ratio of the A$_{1g}$-peak in MoS$_2$ and G-peak in QFEG (Fig. 4.4c) clearly indicates that the defect level is higher along QFEG wrinkles and in locations where the Raman spectra indicate bulk MoS$_2$ (Fig. 4.4c,e).

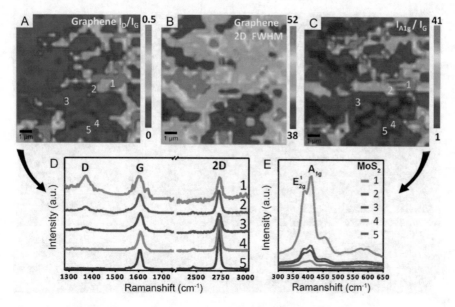

Fig. 4.4 (**a**) The Raman ratio map of QFEG I_D/I_G indicates that the synthesis of MoS_2 can induce defects in the underlying graphene layers at nucleation sites (i.e., wrinkles, steps). (**b**) Corresponding 2D FWHM map of QFEG shows that broadening of the graphene 2D width can correlate the MoS_2 thickness. (**c**) The Raman peak ratio map of I_{A1g} of MoS_2 to I_G of QFEG ratio indicates the MoS_2 distribution on QFEG, where higher ratios indicate "thicker" MoS_2. Comparing (**a**, **c**), there is a clear correlation between defects and MoS_2 thickness at QFEG wrinkles. (**d**, **e**) Raman spectra of numbered locations in the Raman map presented the MoS_2 and QFEG (**a**–**c**) [29]

Interestingly, every MoS_2 island associated with a wrinkle appears to be bisected by the wrinkle shown in Fig. 4.2b, suggesting that the nucleation of the MoS_2 occurs at the peak of the wrinkle and is followed by lateral growth of the layer. This scenario is highly plausible considering that the wrinkle apex is of highest stress and thus the highest chemical reactivity [29]. It has been reported that chemical functionalization on graphene is energetically preferable to happen at the site with higher chemical reactivity due to a lower formation energy [29]. Thus, we speculate that the formation energy of MoS_2 nucleation is lower at the location with high chemical reactivity.

The direct synthesis of TMD on QFEG results in a pristine hetero-interface. Transition electron microscopy (TEM) cross-sectional micrographs confirm that monolayer MoS_2 nucleates on the $(1\bar{1}0n)$ step edges and subsequently extends onto (0001) terrace in regions where the thickness of graphene doesn't vary over the $(1\bar{1}0n)/(0001)$ conjunction (Fig. 4.5a). This suggests that the $(1\bar{1}0n)$ step edges may play a critical role in nucleation, which is consistent with the observation in AFM images of $MoS_2/EG(QFEG)$ in (a, b) of Fig. 4.2. Furthermore, the MoS_2 layer appears to be "blind" to thickness variations in the underlying graphene when there are no defects in the top layer of the graphene (Fig. 4.5b).

Apparently, in the case where additional layers of graphene are formed in a manner as to maintain a flat surface profile, the graphene appears to shield the

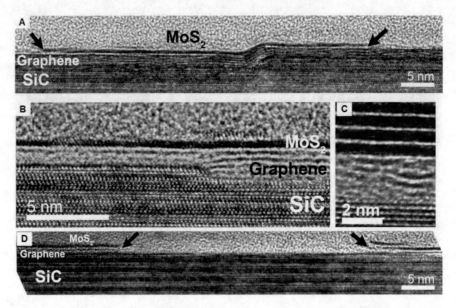

Fig. 4.5 (a) Cross-sectional HRTEM of MoS₂/QFEG demonstrating the nucleation and subsequent lateral growth of MoS₂ on a SiC step edge covered with QFEG. In this case, the graphene thickness is consistent across the step, resulting in bending of the graphene and variation in strain of the EG layer. On the other hand, when the top graphene layer remains flat (**b**), the MoS₂ grows without regard of the changing morphology below. When the underlying graphene is defective (**c**), additional MoS₂ adlayers are present, which indicate defects in the graphene can produce this MoS₂. Finally, MoS₂ nucleation and growth is promoted by the presence of graphene, while often is found to be absent on bare SiC [29]

influence from the SiC morphology. On the other hand, when defective graphene is present at the surface of the graphene layer, there is almost always multilayered MoS₂ present (Fig. 4.5c). Additionally, quite often it is found that MoS₂ nucleation and growth does not proceed on the SiC (0001) terrace when graphene is not present. Figure 4.5d demonstrates that the MoS₂ growth ends at the terminations of graphene, which suggests that graphene may serve as the catalytic layer for MoS₂ nucleation and growth. The similar case has been reported in the growth of MoS₂ on CVD graphene/Cu [15].

Photoconductivity has been demonstrated on monolayer MoS₂, but the photoconductivity must be improved to be competitive to current state-of-the-art materials. Yin et al. [35] demonstrate a photoresponsivity of ~0.42 mA/W with at a photon intensity of 80 μW and drain bias of 1 V, while 7.5 mA/W is achieved at $V_g = 50$ V. This was improved upon by Zhang et al. [39] with the mechanically stacked heterostructures made of synthetic MoS₂/graphene, where the photoresponsivity reached more than 10^7 A/W with the electrically gating effect. We also investigate photoconductivity and responsivity using two terminal MoS₂/QFEG devices without back-gating fabricated via standard photolithography process to demonstrate the superiority of direct synthesis heterostructures [24]. The channel

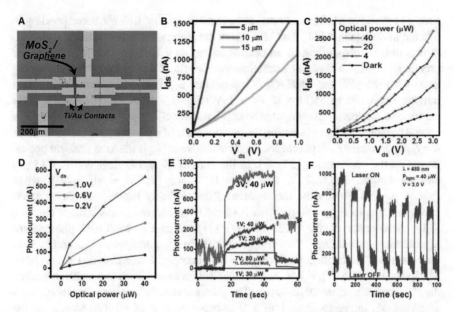

Fig. 4.6 (a) Optical image of the fabricated MoS$_2$/graphene photosensor. (b) Room-temperature electrical characteristic of a two-terminal MoS$_2$/QFEG photosensor shows the drain current (I_{ds}) vs voltage (V_{ds}) at different channel lengths (5,10,15 µm). The output characteristics of the MoS$_2$/QFEG photosensor (**c,d**) when illuminated with 488 nm photon at increasing illuminating laser power and gate voltage. (**e**) Photoswitching behavior of the photosensor at different laser power (P_{light}) and V_{ds} compared to the literature [35] and (**f**) transient measurement of the photosensor at V_{ds} at 3.0 V and P_{light} at 40 µW demonstrating fast, stable response [29]

resistance is found to increase from hundreds of Ohms to ~10^6 Ohms as the channel length increases from 5 to 15 µm (Fig. 4.6a,b). Interestingly, the drain current is similar to previous reports for MoS$_2$ [35], and much lower than epitaxial graphene devices [22], indicating that the Ti/Au metal contacts have not contacted the underlying graphene and that transport is dominated by MoS$_2$ in the channel. This is encouraging for the development of lateral TMD devices on graphene, since the graphene has not shorted the device. This phenomenon is likely a result of the high interlayer resistance in vdW solids [40], which provides a large potential barrier for vertical transport and results in charge being confined to the top MoS$_2$ layers when the ohmic metallization does not contact the underlying graphene. In the case where pinholes exist in the MoS$_2$ layer under the ohmic metallization, we find that current transport is dominated by graphene, and *I–V* curves become highly linear with channel resistances <500 Ohms. These devices were not considered for photocurrent measurements. Direct growth of MoS$_2$ on EG yields significant improvement in photoconductivity and responsivity. The MoS$_2$/QFEG photosensor device was investigated by exploring photocurrent under various optical powers and bias conditions. Ideally, the generation of photocurrent needs to match the basic condition that the incident photon energy must be greater than the optical energy gap (E_g) of

MoS_2. Monolayer MoS_2 is reported to have a bandgap of 1.83 eV corresponding to 676 nm in wavelength [34]. The MoS_2 from the direct growth in this study shows PL at 668 nm, corresponding to an optical energy gap 1.85 eV that was confirmed preciously. Under a constant excitation wavelength at 488 nm, and power ranging from 4 to 40 µW, the MoS_2/QFEG generates a power-dependent photocurrent ranging from 150 to 550 nA at V_{ds} = 1 V (Fig. 4.6c,d). This represents a 200× increase in photocurrent compared to exfoliated MoS_2 [35]. Photoresponsivity is a crucial parameter when evaluating the performance of a photosensor. It is defined as $R = I_{ph}/P_{light}$, where I_{ph} is photogenerated current and P_{light} is the total incident optical power on the photosensor [35, 39]. In the current MoS_2/EG device with a 15 µm channel length, V_{ds} = 1 V, and V_g = 0 V, we find R = 40 mA/W when P_{light} = 40 µW. Previously, the reported R of similarly biased monolayer MoS_2 photosensor with P_{light} = 80 µW was 0.42 mA/W [35], which was 2 times that of a WS_2 photosensor device (0.22 mA/W) [41]. Thus the MoS_2/EG photosensor represents a minimum 100× improvement in photoresponsivity, and 1000× improvement in absolute photoconductivity, compared to exfoliated single-layer MoS_2 (Fig. 4.6e). At V_{ds} = 1 V the photocurrent increases from 150 to 230 nA when the P_{light} increases from 20 to 40 µW. The photocurrent further increases to 1000 nA when the V_{ds} increases from 1 to 3 V at constant P_{light} of 40 µW. Moreover, the stability of this switching behavior is further tested by sequentially turning the illumination laser on (30 s) and off (30 s). Figure 4.6e shows the photocurrent stability over 1000 s of continuous operation. Correlating materials characterization with optoelectronic performance, the improvement can be ascribed to (1) the high crystal quality of MoS_2 grown on QFEG and the presence of pristine surfaces and atomically sharp interfaces of the MoS_2/QFEG heterostructures and (2) the underlying QFEG screens charge scattering sites that may be present in the SiC substrate.

In order to determine the universality of epitaxial graphene as a template for the synthesis of high-quality vdW solids, we further demonstrate the integration of WSe_2 and hBN on QFEG substrates. Cross-sectional HRTEM of WSe_2/QFEG/SiC and Raman of hBN/QFEG/SiC (Fig. 4.7) indicate that the underlying graphene is not damaged by the synthesis process [2]. Additionally, TEM confirms a clean vdW gap between the different layered materials. After the growth of hBN on QFEG, several Raman peaks are present at 1368 cm^{-1}, which is suggested as the combination of the E_{2g} mode of B–N vibration at 1372 cm^{-1} and the D peak of graphene at 1360 cm^{-1} [14]. The very minor peak at 2950 cm^{-1} (corresponding to the graphene Raman D + G mode) indicates the growth of hBN only nominally increases the defect density. Additionally, similar to MoS_2, there is no carbon bonding in the W-, Se-, or B-XPS peaks, indicating minimal interaction between QFEG and the overlayer. As a result, the utilization of EG as a template may be considered as an excellent candidate for development of a broad range of vertical heterostructures.

Fig. 4.7 Transmission
electron microscopy of (a)
WSe$_2$/QFEG reveals that
the interface within layers
and between heterolayers
is pristine, with no
observable defects. (b)
Additionally, Raman
spectra of quasi-
freestanding epitaxial
graphene (QFEG) and
QFEG/hBN indicate
high-quality hBN has been
grown on QFEG. There is
also a small PL
background following hBN
synthesis [29]

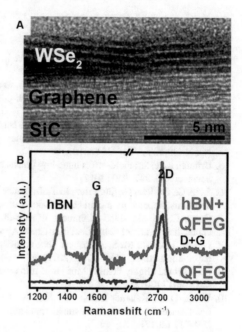

4.4 Conclusions

High-quality EG has been synthesized and utilized as the growth template for vdW
solids that include MoS$_2$, WSe$_2$, and hBN. TMDCs are synthesized by gas-phase
reaction of MoO$_3$ and WO$_3$ and S and Se, while hBN is synthesized on epitaxial
graphene by utilizing ammonia borane as the precursor in a tube CVD chamber.
This demonstrates that EG may be a universal substrate for a variety of deposition
methods and materials. This is due to the presence of residual strain and wrinkles,
which play an important role in the nucleation of vdW solids. It has been found that
QFEG template has less MoS$_2$ nucleated on terrace than EG template due to the
strain relief and elimination of the underlying buffer – both expected to increase
chemical reactivity of the graphene. Photosensors based on direct growth MoS$_2$/
QFEG heterostructures exhibit an improvement in photoresponsivity by a minimum
of 200×, demonstrating the high quality in heterostructures from the direct growth.
Finally, we have provided strong evidence that EG can be an excellent candidate for
building large-area vdW solids that will have extraordinary properties and
performance. The two-terminal electrical measurements on synthesized layers here
show that the interlayer resistances between the graphene and top heterolayers are
high enough and that the stacks are not "shorted" simply by the presence of
graphene, rather it will be proper contacting of the individual layers that will be
critically important.

References

1. Novoselov, K.S., et al.: Electric field effect in atomically thin carbon films. Science. **306**, 666–669 (2004)
2. Bresnehan, M.S., et al.: Integration of hexagonal boron nitride with quasi-freestanding epitaxial graphene: toward wafer-scale, high-performance devices. ACS Nano. **6**, 5234–5241 (2012)
3. Geim, A.K., Grigorieva, I.V.: Van der Waals heterostructures. Nature. **499**, 419–425 (2013)
4. Dean, C.R., et al.: Boron nitride substrates for high-quality graphene electronics. Nat. Nanotechnol. **5**, 722–726 (2010)
5. Britnell, L., et al.: Field-effect tunneling transistor based on vertical graphene heterostructures. Science. **335**, 947–950 (2012)
6. Sanz, C., Guillén, C., Gutiérrez, M.T.: Influence of the synthesis conditions on gallium sulfide thin films prepared by modulated flux deposition. J. Phys. D. Appl. Phys. **42**, 085108 (2009)
7. Gong, C., et al.: Band alignment of two-dimensional transition metal dichalcogenides: application in tunnel field effect transistors. Appl. Phys. Lett. **103**, 053513 (2013)
8. Lee, H.S., et al.: MoS_2 nanosheet phototransistors with thickness-modulated optical energy gap. Nano Lett. **12**, 3695–3700 (2012). https://doi.org/10.1021/nl301485q
9. Pu, J., et al.: Highly flexible MoS_2 thin-film transistors with ion gel dielectrics. Nano Lett. **12**, 4013–4017 (2012)
10. Wang, Q.H., Kalantar-Zadeh, K., Kis, A., Coleman, J.N., Strano, M.S.: Electronics and optoelectronics of two-dimensional transition metal dichalcogenides. Nat. Nanotechnol. **7**, 699–712 (2012)
11. Wang, J.-Z., et al.: Development of MoS_2-CNT composite thin film from layered MoS_2 for lithium batteries. Adv. Energy Mater. **3**, 798–805 (2013)
12. Yan, K., Peng, H., Zhou, Y., Li, H., Liu, Z.: Formation of bilayer bernal graphene: layer-by-layer epitaxy via chemical vapor deposition. Nano Lett. **11**, 1106–1110 (2011)
13. Dang, W., Peng, H., Li, H., Wang, P., Liu, Z.: Epitaxial heterostructures of ultrathin topological insulator nanoplate and graphene. Nano Lett. **10**, 2870–2876 (2010)
14. Liu, Z., et al.: Direct growth of graphene/hexagonal boron nitride stacked layers. Nano Lett. **11**, 2032–2037 (2011)
15. Shi, Y., et al.: Van der Waals epitaxy of MoS_2 layers using graphene as growth templates. Nano Lett. **12**, 2784–2791 (2012)
16. Radisavljevic, B., Radenovic, A., Brivio, J., Giacometti, V., Kis, A.: Single-layer MoS_2 transistors. Nat. Nanotechnol. **6**, 147–150 (2011)
17. Kubota, Y., Watanabe, K., Tsuda, O., Taniguchi, T.: Deep ultraviolet light-emitting hexagonal boron nitride synthesized at atmospheric pressure. Science. **317**, 932–934 (2007)
18. Gorbachev, R.V., et al.: Strong coulomb drag and broken symmetry in double-layer graphene. Nat. Phys. **8**, 896–901 (2012)
19. Haigh, S.J., et al.: Cross-sectional imaging of individual layers and buried interfaces of graphene-based heterostructures and superlattices. Nat. Mater. **11**, 764–767 (2012)
20. Riedl, C., Coletti, C., Iwasaki, T., Zakharov, A.A., Starke, U.: Quasi-free-standing epitaxial graphene on SiC obtained by hydrogen intercalation. Phys. Rev. Lett. **103**, 246804 (2009)
21. Lin, Y.-C., et al.: Graphene annealing: how clean can it be? Nano Lett. **12**, 414–419 (2012)
22. Robinson, J.A., et al.: Epitaxial graphene transistors: enhancing performance via hydrogen intercalation. Nano Lett. **11**, 3875–3880 (2011)
23. Perea-López, N., et al.: Photosensor device based on few-layered WS_2 films. Adv. Funct. Mater. **23**, 5511–5517 (2013)
24. Robinson, J.A., et al.: Contacting graphene. Appl. Phys. Lett. **98**, 053103 (2011)
25. Robinson, J.A., Puls, C.P., Staley, N.E., Stitt, J.P., Fanton, M.A.: Raman topography and strain uniformity of large-area epitaxial graphene. Nano Lett. **9**, 964–968 (2009)
26. Wilson, J.A., Yoffe, A.D.: The transition metal dichalcogenides discussion and interpretation of the observed optical, electrical and structural properties. Adv. Phys. **18**, 193–335 (1969)

27. Koma, A.: Van der Waals epitaxy – a new epitaxial growth method for a highly lattice-mismatched system. Thin Solid Films. **216**, 72–76 (1992)
28. Röhrl, J., et al.: Raman spectra of epitaxial graphene on SiC(0001). Appl. Phys. Lett. **92**, 201918 (2008)
29. Lin, Y.-C., et al.: Direct synthesis of van der Waals solids. ACS Nano. **8**, 3715–3723 (2014)
30. Choi, J.S., et al.: Friction anisotropy-driven domain imaging on exfoliated monolayer graphene. Science. **333**, 607–610 (2011)
31. Bissett, M.A., Konabe, S., Okada, S., Tsuji, M., Ago, H.: Enhanced chemical reactivity of graphene induced by mechanical strain. ACS Nano. **7**, 10335–10343 (2013)
32. Lee, C., et al.: Anomalous lattice vibrations of single- and few-layer MoS_2. ACS Nano. **4**, 2695–2700 (2010)
33. Gong, C., et al.: Metal contacts on physical vapor deposited monolayer MoS_2. ACS Nano. **7**, 11350–11357 (2013)
34. Baskaran, A., Smereka, P.: Mechanisms of Stranski-Krastanov growth. J. Appl. Phys. **111**, 044321 (2012)
35. Yin, Z., et al.: Single-layer MoS_2 phototransistors. ACS Nano. **6**, 74–80 (2012)
36. Tang, S., et al.: Nucleation and growth of single crystal graphene on hexagonal boron nitride. Carbon. **50**, 329–331 (2012)
37. Ki K., K., et al.: Enhancing the conductivity of transparent graphene films via doping. Nanotechnology **21**, 285205 (2010)
38. Li, Q., et al.: Polycrystalline molybdenum disulfide ($2H-MoS_2$) Nano- and Microribbons by electrochemical/chemical synthesis. Nano Lett. **4**, 277–281 (2004)
39. Zhang, W., et al.: Ultrahigh-Gain Photodetectors Based on Atomically Thin Graphene-MoS_2 Heterostructures. Sci. Rep. **4**, 3826 (2014)
40. Das, S., Appenzeller, J.: Screening and interlayer coupling in multilayer MoS_2. Phys. Status Solidi RRL **7**, 168–273 (2013)
41. Perea-López, N. et al.: Photosensor Device Based on Few-Layered WS_2 Films Adv. Func. Mater. **23**, 5511–5517 (2013)

Chapter 5
Atomically Thin Heterostructures Based on Monolayer WSe$_2$ and Graphene

5.1 Introduction

Heterogeneous engineering of two-dimensional layered materials, including metallic graphene and semiconducting transition metal dichalcogenides, presents an exciting opportunity to produce highly tunable electronic and optoelectronic systems. In order to engineer pristine layers and their interfaces, epitaxial growth of such heterostructures is desirable. We report the direct growth of crystalline and monolayer tungsten diselenide (WSe$_2$) on epitaxial graphene (EG) on silicon carbide. Raman spectroscopy, photoluminescence, and scanning tunneling microscopy confirm high-quality WSe$_2$ monolayers, while transmission electron microscopy shows an atomically sharp interface, and low-energy electron diffraction confirms near-perfect orientation between WSe$_2$ and EG. Vertical transport measurements across the WSe$_2$/EG heterostructure provide evidence that an additional barrier to carrier transport beyond the expected WSe$_2$/EG band offset exists due to the interlayer gap, which is supported by theoretical local density of states (LDOS) calculations using self-consistent density functional theory (DFT) and non-equilibrium Green's function (NEGF).

Analogous to the evolution in graphene research [1], the research community is at the initial stage of forming and characterizing vdW heterostructures, where samples are produced mainly through mechanical exfoliation and manual transfer stacking [2]. Unlike isolated monolayer samples, the transfer stacking process can lead to uncontrollable interface contamination [3] that in turn results in reduced device performance [2, 3]. Therefore, developing synthetic techniques to form such heterostructures is critical for engineering pristine layers and junction interfaces. Efforts toward this end include the vertical integration of 2D materials such as MoS$_2$ and hBN on EG [4]. Similarly, CVD graphene grown on Cu foils has been utilized as "universal template" for the synthesis of vertical hBN or MoS$_2$ [5, 6] or lateral (in-plane) hBN/graphene systems [7]. In either case, monolayer growth control is

© Springer Nature Switzerland AG 2018
Y. -C. Lin, *Properties of Synthetic Two-Dimensional Materials and Heterostructures*, Springer Theses,
https://doi.org/10.1007/978-3-030-00332-6_5

essential to exploit phenomena such as the direct-gap crossover in TMDC [8] or interlayer coupling that can hybridize the electronic structure of stacked monolayers [9]. In this report, we demonstrate direct growth of high-quality WSe$_2$ monolayers on EG and provide evidence that this heterosystem exhibits pristine interfaces; high-quality structural, chemical, and optical properties; and significant tunnel resistances due to the WSe$_2$/EG interlayer gap.

5.2   Experimental Methods

5.2.1   Growth and Properties of WSe$_2$ Layers on Graphene

Epitaxial graphene is grown on diced SiC wafers via sublimation of silicon from 6H-SiC (0001) at 1725 °C for 20 min under 200 Torr argon (Ar) background pressure after hydrogen (H$_2$) etching at 1500 °C for 15 min under 700 Torr 10% H$_2$/Ar mixtures.

Tungsten trioxide (WO$_3$) powder (0.3 g) was placed in a ceramic crucible located in the center of the furnace. The Se powder is placed in a separate ceramic boat upstream of the WO$_3$ and maintained at 270 °C during the reaction. Graphene substrates were subsequently placed downstream of the crucible loaded with WO$_3$, and Se and WO$_3$ vapors were transported to the targeting sapphire substrates by an Ar/ H$_2$ (80:20) forming gas at 1 Torr of chamber pressure. The center of the hot zone was held at 925 °C for 30 min to 1 h, and the furnace was then naturally cooled to room temperature. The as-grown heterostructures are characterized using atomic force microscopy (AFM), Raman spectroscopy, transmission electron microscopy (TEM), and X-ray photoelectron spectroscopy (XPS). A WITec CRM200 confocal Raman microscope with an excitation wavelength (488 nm) is utilized for structural characterization. A Bruker Dimension with a scan rate of 0.5 Hz was utilized for the AFM measurements. The TEM cross-sectional samples were made via utilizing a NanoLab dual-beam FIB/SEM system. Protective layers of SiO$_2$ and Pt were deposited to protect the interesting region during focused ion beam milling. TEM imaging was performed using a JEOL 2100F operated at 200 kV. CAFM measurement was performed in Bruker Dimension. For surface analysis, the sample was loaded into an ultrahigh vacuum (UHV) with a base pressure in low 10^{-10} mbar range described in detail elsewhere [10]. The WSe$_2$/EG sample was then imaged using an Omicron variable temperature scanning tunneling microscope (STM) without any thermal treatment. The STM images were obtained at room temperature and in the constant current mode, with an etched tungsten tip. The same system is equipped with a monochromatic Al-Kα source (E = 1486.7 eV) and an Omicron Argus detector operating with pass energy of 15 eV. The spot size used during the acquisition is equal to 0.5 mm. Core-level spectra taken with 15 sweeps are analyzed with the spectral analysis software analyzer [11].

5.2.2 Diode Fabrication

The heterojunction device demonstrated is fabricated with electron-beam lithography (EBL) and lift-off of evaporated metal contacts. The process flow is illustrated in Fig. 5.1. In the first step, the graphene contact is patterned and developed with EBL. Subsequently, metal contacts Ti/Au (10 nm/40 nm) are deposited with low-pressure electron beam evaporation (10^{-7} Torr) after an oxygen plasma treatment to reduce the contact resistance (45 s at 100 W, 50 sccm He, 150 sccm O_2 at 500 mTorr) [13]. Then a layer of 30 nm Al_2O_3 is deposited conformally over the entire substrate with atomic layer deposition (ALD), which serves as a protection layer for subsequent processing steps and a passivation layer. ALD-deposited Al_2O_3 capping layer has been reported as an effective film to substantially improve carrier mobility in 2D materials [14]. In the second EBL step, a pattern of etched regions are defined, including an opening on the Ti/Au pads, and regions for the later WSe_2 contacts. The Al_2O_3 capping layers on these regions are first removed with hydrofluoric acid, followed by oxygen plasma etching to remove the monolayer WSe_2 and few layers of graphene. This step prevents shorting through the underlying graphene layer after depositing the WSe_2 contacts. In the third e-beam step, the WSe_2 contact pads and thin lines are defined, and the Al_2O_3 layer on the WSe_2 triangular sheets is removed by hydrofluoric acid prior to the metal deposition. Then 50 nm thick Palladium (Pd) layer is deposited by electron beam evaporation at 10^{-7} Torr. The high work function Pd contacts with WSe_2 have been reported to produce a smaller Schottky barrier, and many orders have higher current density compared to Ti/Au contacts [15].

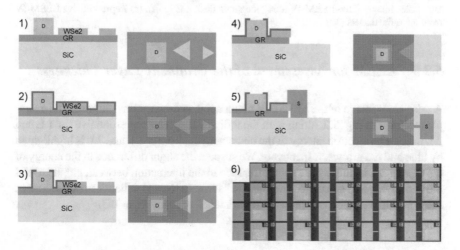

Fig. 5.1 Flow for the fabrication of diodes on WSe_2-graphene: (1) patterning and depositing Ti/Au for graphene contacts. (2) Coating the entire substrate with ALD Al_2O_3 (present as the dark mesh layer). (3) Define the isolation region on graphene contact pads, and later WSe_2 contact regions, and remove the Al_2O_3 layer. (4) Isolation etching with oxygen plasma to remove the WSe_2 and graphene. (5) Patterning the WSe_2 contacts (Pd). (6) Finished arrays of diode devices on SiC [12]

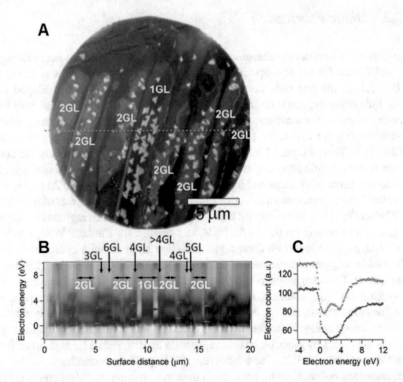

Fig. 5.2 (**a**) LEEM image acquired at the electron energy 1.6 eV above the vacuum level. (**b**) The false color image of the LEEM-IV along the green dash line in (**a**). (**c**) Representative LEEM-IV from the same dataset [12]

5.2.3   LEEM for Assessment of the Graphene Layer Thickness

A false color image of the LEEM-IV along a line is used for this purpose. The example is shown in Fig. 5.2, along with the individual spectra representative for 1 L and 2 L in (**c**). The electron energy of the characteristic dips for 1 L and 2 L is highlighted by blue and red arrows, respectively. We suspect the slight difference in the energy of the dips in 2 L is due to the slight difference in the interaction between graphene and the underlying SiC substrate. The "narrow" terraces located in between large terraces show multiple dips consistent with the existence of many-layer EG.

5.3   Results and Discussion

Epitaxial graphene grown from silicon carbide (SiC) [16, 17] is an ideal platform to investigate the nucleation and growth of vdW heterostructures. In particular, EG on SiC eliminates the need for post-growth transfer required for chemical vapor-deposited graphene and therefore provides a chemically uniform starting surface.

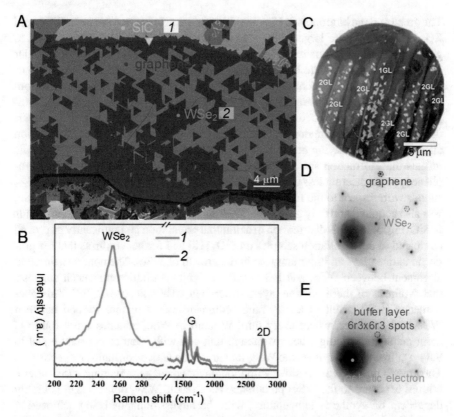

Fig. 5.3 (a) SEM image of WSe₂ monolayers grown on epitaxial graphene shows the WSe₂ triangular crystals strictly aligned on graphene due to their commensurability. (**b**) Evident by Raman spectroscopic measurement, WSe₂ monolayers only grew on the area where graphene exists. (**c**) LEED pattern of selected circles with a diameter of 5 μm was obtained and shows zero misorientation between WSe₂ and graphene in (**d**) and the presence of buffer layer only on SiC in (**e**) [12]

Epitaxial graphene also enables the nucleation and epitaxial growth of WSe₂. Scanning electron microscopic image shows the selective growth of WSe₂ on graphene which indicates that a significant surface energy difference between SiC and EG exists and that EG is more favorable for nucleation and growth of the WSe₂ (Fig. 5.3a). Raman spectra show that the WSe₂ (A_{1g}/E_{2g} peak ~250 cm^{-1}) only grows where the graphene (1580 cm^{-1} and 2700 cm^{-1}) presents (Fig. 5.3b). This phenomenon may be quite useful for developing templated growth of vdW heterostructures in the future and requires subsequent theory to better understand the fundamental mechanisms of growth selectivity. The WSe₂ triangular domains seem to strictly align to the underlying graphene. Using low-energy electron microscopy/diffraction (LEEM/LEED), we are able to more precisely quantify the in-plane crystalline orientation of WSe₂ monolayers on EG (Fig. 5.3c–e). Unlike the diffuse LEED patterns of monolayer MoS₂ or MoSe₂ on SiO₂ [18, 19], the LEED spots of monolayer WSe₂/EG are sharp, resembling that found for twisted bilayer graphene on SiC [20].

The green and the blue spots in Fig. 5.3d illustrate the diffraction spots for WSe$_2$ and EG, respectively. The larger WSe$_2$ lattice constant (3.28 Å) [21] as compared to graphene (2.46 Å) means the WSe$_2$ diffraction spots will be closer to the specular beam (central spot), where both crystals display hexagonal symmetry. The ratio of their lattice constants matches the ratio of the hexagons' sizes (~1.3, extracted from our experiment), demonstrating a 23% lattice mismatch between WSe$_2$ and graphene. Acquiring diffraction from multiple WSe$_2$ islands simultaneously reveals that the WSe$_2$ is not randomly orientated, but maintains an in-plane orientation aligned to the underlying graphene layer [12]. A detailed inspection of the sample reveals that diffraction spots of WSe$_2$ show near-perfect alignment with the graphene spots, indicating less than ±5° variation of the relative orientation of the WSe$_2$ islands with respect to the EG layer (not shown). The area without as-grown WSe$_2$ was also acquired and only gave the diffraction spot from the buffer layer that bonds to SiC (Fig. 5.3e). This distribution of azimuthal orientation is significantly improved compared to the exfoliation samples on SiO$_2$ [18] and further confirms that the graphene lattice plays an important role in the growth of WSe$_2$. The notable azimuthal alignment between WSe$_2$ and EG despite a significant lattice mismatch suggests that synthesis of these pristine layers occurs via vdW epitaxy [22, 23]. This phenomenon is the result of a long-range commensurate structure formed between WSe$_2$ and graphene, where every third W atom in WSe$_2$ matches every fourth C atom graphene forming a heterostructure unit cell with a lattice constant equal to 9.84 Å (three and four times the WSe$_2$ and graphene lattice constants, respectively). The growth rate of WSe$_2$ is slow, with monolayer coverage remaining <75% after a 60-min exposure to the WSe$_2$ precursors. As a result, WSe$_2$ domain size is highly dependent on synthesis temperature, with the largest domains being achieved at 1000 °C. Morphological features in the EG (such as wrinkles, SiC step edges, and other surface imperfections) appear to directly influence WSe$_2$ monolayer development by acting as a barrier to further lateral growth or by modifying the registry of the WSe$_2$ layers on epitaxial graphene. A qualitative assessment of the WSe$_2$ in-plane orientation via AFM suggests a narrow distribution with >80% of the triangles aligned (±5°) to the underlying graphene.

It is well known that the electron reflectivity spectra obtained through LEEM measurements (LEEM-IV) can provide the "fingerprint" of the EG thickness [24, 25]. By combining the information of EG's thickness together with the density and size of WSe$_2$ islands, we observe a distinct correlation between WSe$_2$ island density and EG thickness: the terrace areas of SiC terminated with 1–2 layer EG have a higher density of larger WSe$_2$ islands. Electron reflectivity spectra reveal that this particular terrace is terminated with a monolayer of EG, while the neighboring terraces have bilayer EG. We also note that there are virtually no WSe$_2$ islands on the narrow terraces located between the larger terraces or near the step bunches originated from the morphology of the SiC substrate. LEEM-IV confirms these "narrow" terraces are covered by many-layer EG, typically three to five layers. As a result, one must take care to control the layer thickness of the underlying graphene to one to two layers since the surface properties and chemical reactivity of many-layer graphene layers preclude the formation of large WSe$_2$ domains, similar to that found for synthesis on graphite [26]. Cross-sectional TEM reveals that the uniformity and structure of the underlying graphene can significantly impact the nucleation, growth, and quality of the WSe_2 overlayer. Wherever pristine graphene is present, the WSe_2 overlayer is crystalline, with no observable defects (Fig. 5.4a). Additionally, the EG interlayer distance is measured to be 3.64 Å (typical for EG/ SiC) [27], while the WSe_2/EG layer spacing is 5.23 Å, with a WSe_2 thickness of 6.45 Å [21]. In the case where graphene is defective, we also observe increased disorder in the WSe_2. Unlike the MoS_2 grown on EG [4], in which the SiC ($1\bar{1}0n$) step edges and EG wrinkles serve as nucleation sites for MoS_2 growth, the WSe_2 abruptly stops at the edge of the ($1\bar{1}0n$) plane (also see STM in Fig. 5.4b), preferring to grow only on EG synthesized on the SiC (0001) plane. It is also useful to note that WSe_2 is sensitive to damage during TEM investigations and it is not stable under high-energy electrons (also seen in LEEM at high electron beam intensity). The height of individual WSe_2 domains measures 0.71 nm (Fig. 5.4b), and atomic arrangement matches that of $2H$-WSe_2 [28], in good agreement with previous reports on CVD WSe_2 [29]. The bandgap size of the WSe_2 monolayer was measured in scanning tunneling spectroscopic measurement.

The STS profile shows 1.87 eV of the 1 L WSe_2 bandgap and a parabolic curve of conductivity that belongs to few-layer graphene (Fig. 5.4c). The Fermi level of WSe_2 is at the middle of bandgap, which means that its transport will not favor either p- or n-type. To better understand the impact of the graphene under the WSe_2, we compare the PL spectra of WSe_2 grown on insulating sapphire formed under the same growth conditions (Fig. 5.4d). From this comparison, three features are apparent: (i) the PL intensity of WSe_2/EG is quenched by a factor of three; (ii) the PL peak position of WSe_2/EG is upshifted by 35 meV (from 1.625 eV on sapphire to 1.66 eV on EG); and (iii) the full width of half-maximum (FWHM) of the WSe_2 PL peak on EG is narrower than WSe_2-on-sapphire (38 meV versus 80 meV). Since the WSe_2 crystalline quality is known to be high on EG based on the characterization results, the PL quenching is likely a result of photo-generated charge carriers transferring from WSe_2 to EG [24, 25]. Zhang et al. [30] proposed the observed quenching in MoS_2/Gr is due to the exciton splitting by the built-in electrical field between 1 L CVD graphene and 1 L CVD MoS_2. Additionally, Shim et al. [31] observed quenching in $MoSe_2$/graphene heterostructures due to a fast, non-radiative recombination process. It is likely that the quenching process for WSe_2/EG observed here is similar in nature, especially since there is p-type doping in WSe_2 when in proximity to EG. The second feature (red shift of PL peak) may be the result of strain, doping, or defects [31]. Within the resolution of our measurements (LEED, STM, and PL), we find the WSe_2 defect density to be low; however, there is measurable doping of the WSe_2 as a result of the underlying EG (based on XPS results). Doping is known to shift PL signatures in MoS_2 [32], and we believe it is the primary contributor to the measured PL shift observed in this work. Finally, the PL properties of WSe_2 on an insulating substrate (sapphire) versus on graphene are very similar to a recent report comparing in MoS_2/SiO_2 versus MoS_2/EG and MoS_2/hBN [33]. The narrower peak width suggests that the interface between the WSe_2 and EG is pristine, with no dangling bonds contributing to interface roughness or surface optical phonon scattering that in turn leads to an improved optoelectronic quality.

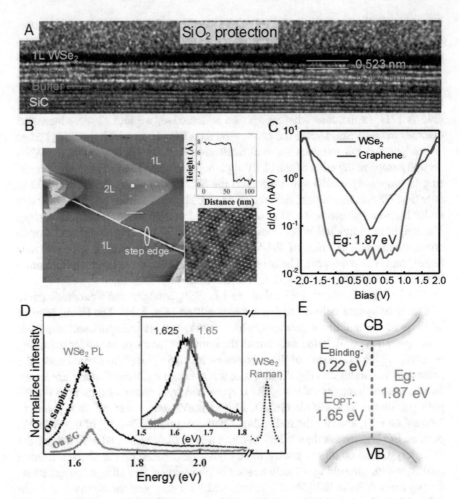

Fig. 5.4 (a) Cross-sectional HRTEM image of WSe$_2$-EG shows the false-colored constituent layers and the separation distance of WSe$_2$-graphene and graphene-graphene. (b) AFM image shows a full coverage of 1 L WSe$_2$ on graphene. Two liter and thicker WSe$_2$ started to form along the step edge of SiC due to its higher surface energy. The thickness of a layer is 6.5 nm. High-resolution STM image shows the atomic structure of the WSe$_2$ lattice in a hexagonal pattern. (0.35 V, 1.5 nA). (c) STS profiles show gapless graphene and 1.87 eV of the WSe$_2$ bandgap in room temperatures. (d) PL spectra normalized to the WSe$_2$ Raman peak show the PL intensity quenches on graphene due to the metallic nature of graphene. Inset is the normalized PL peaks and shows their peak positions. (e) The relationship between the optical and electrical bandgap of WSe$_2$ was tightened by a binding energy [12]

There is a difference in energy (220 meV) between the measured optical bandgap and electrical bandgap (Fig. 5.4e). The electronic bandgap characterizes single-particle or quasiparticle excitations and is defined by the sum of the energies needed to separately tunnel an electron and a hole into monolayer TMDC. On the other hand, the optical bandgap is described as the energy required to create an exciton, which is correlated to two-particle electron-hole pair, through optical absorption.

The difference in these two energies directly yields the exciton binding energy (E_{Binding}). This exciton binding energy is one order larger than those of traditional semiconductors due to enhanced Coulomb interactions and due to low-dimensional effects which are expected to increase the quasiparticle bandgap as well as to cause electron-hole pairs to form more strongly bound excitons.

Synthesis of WSe_2/EG results in pristine chemical, optical, and structural quality of the heterostructure layers and interfaces. Raman and X-ray photoelectron spectroscopy (XPS) confirm there is no measureable reaction between graphene and WSe_2, and the integrals of high-resolution spectra of the Se 3d and W 4f peaks lead to an estimated Se:W ratio of ~2:1 [12]. Recent reports of core-level energies of monolayer WSe_2 on an insulating sapphire substrate using non-monochromatic Mg Kα X-rays are noted (W4f$_{7/2}$ and W4f$_{5/2}$ peaks are at 32.8 eV and 35.0 eV, respectively; Se3d$_{5/2}$ and 3d$_{3/2}$ peaks are at 55.0 eV and 55.9 eV, respectively) [29]. The study of bulk, exfoliated p-type WSe_2 with monochromatic AlKα$_1$ x-rays indicates that these peaks are shifted to lower values by approximately 0.1 eV [28]. Under identical analysis conditions and parameters to that employed by McDonnell et al. [28], we find here that the monolayer WSe_2/EG exhibits similar binding energies to the exfoliated bulk. Based on the core level measurements of the WSe_2 bulk crystal surface that confirm p-type doping [28], these energies are therefore representative of p-type monolayer WSe_2 interfacing with graphene. We also note that a shift in binding energy toward lower energies is consistent with a lower electron density in WSe_2/EG.

In other words, EG withdraws electrons from WSe_2 monolayer, leading to p-doped behavior in the WSe_2 layer [28]. This is confirmed via direct measurements of the occupied valance energy states by XPS showing that the Fermi level is positioned at 0.72 eV, which is 0.11 eV smaller than the mid-gap energy level in 1 L WSe_2 (0.83 eV). Similar shifts have been reported for other thin films on graphene recently (Fig. 5.5a) [34]. Based on the Fermi level position measured in XPS and the optical bandgap and the reported Fermi level position of as-grown EG, a proposed band alignment is illustrated in Fig. 5.5b. Due to their significant difference in carrier concentration, a large Schottky barrier is expected to form at EG/WSe_2 junction. Conductive atomic force microscopy (C-AFM) and vertical diode structures shown later provide a direct means to probe the nanoscale electrical properties of WSe_2/graphene heterostructures and assist in identifying the utility of these materials for advanced electronic and optoelectronic architectures. Comparing AFM surface topography and conductivity acquired at $V_{\text{bias}} = 0.1$ V (Fig. 5.5c,d) in C-AFM clearly indicates that a barrier to transport exists in the heterojunction regions. The mapping also reveals that the WSe_2 is uniformly resistive, while low resistance contact is possible on the graphene layer, with EG wrinkles and SiC step edges exhibiting enhanced conduction through the AFM tip.

Illustrated by the DFT effective potential profile (Fig. 5.6a), the WSe_2 and graphene are held together by the van der Waals interaction, and the resulting interlayer gap, d_{ILG}, forms a finite potential barrier between them. Figure 5.6b illustrates the spatially resolved local density of states (LDOS) under zero bias. It shows that, within the interlayer gap, d_{ILG}, there are no LDOS contributing to transport. Further,

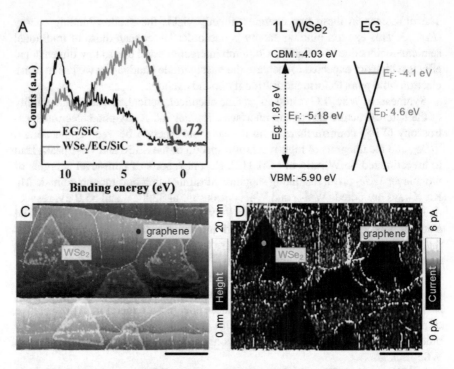

Fig. 5.5 (**a**) XPS measurement confirmed that the valence band maxima is 0.72 eV below the Fermi level. (**b**) Based on the measured spectroscopic results, the band alignment between WSe₂ and graphene (before the contact) was proposed. (**c, d**) The topographic image of as-grown sample and its corresponding C-AFM image (under a bias of 0.1 V) shows WSe₂ is highly resistive due to the presence of electrical barrier between WSe₂ and graphene [12]

the projected contour line of LDOS that delineates the boundary between the negligible (close to zero) LDOS and the finite (0.02) LDOS values provides a quantitative estimate of the transport barrier height arising from the interlayer gap. We estimate this additional barrier to be 1.85 eV above the Fermi level. While this model utilizes intrinsic WSe₂ with pristine contacts, it highlights that the gap between the layers plays a critical role in the determination of the turn-on voltage of the layer stack. The diode structures fabricated via electron-beam lithography and lift-off process (Fig. 5.6c) confirm the presence of a tunnel barrier to vertical transport with turn-on occurring at > ±1.8 V (Fig. 5.6d). The interlayer gap-related barrier to transport persists up to a bias of 1.80 V, acting like a thermionic barrier, as evidenced by no appearance of LDOS – agreeing well with experimental measurements showing device turn-on at ~1.8–2 V. The interlayer gap barrier starts to collapse at a bias beyond 1.85 V, where LDOS appears and contributes to the transport (see Fig. 5.6b). We find theoretically that the barrier due to the interlayer gap depends upon (a) the interlayer gap thickness, d_{ILG}, and (b) the Coulombic interaction among the different atoms of the constituent layers. With the decrease in the

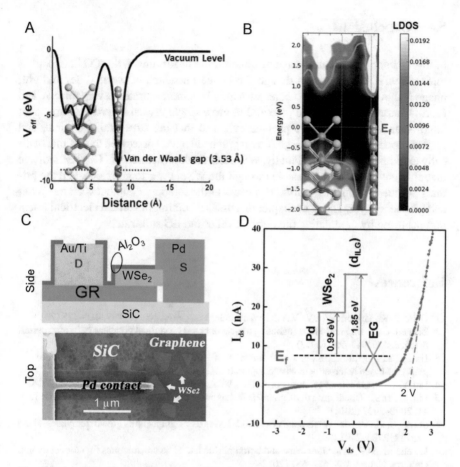

Fig. 5.6 (a) The effective potential profile of a pristine WSe₂/EG heterostructure supercell. Calculated by density functional theory (DFT) along the out-of-plane direction demonstrates that a significant finite barrier to electron transport can exist as a result of the interlayer gap (d_{ILG}). (b) The corresponding local density of states was extracted for the monolayer WSe₂ and the first layer of EG. (c) The side and top view of WSe₂-EG diode that was measured to understand the vertical electrical transport. (d) The measured current versus voltage (I–V) curves from WSe₂/EG diodes confirm a large barrier to transport through the heterostructure as well as ~10^5 on/off ratio and turn-on voltage of ~2 V. Inset shows a schematic different theoretical barrier height with respect to the Fermi level in this device [12]

interlayer gap, the interaction increases which reduces the barrier height. For instance, in our system, we observe that the Pd/WSe₂ interlayer gap distance of 2.98 Å is lower than that of the WSe₂/EG 3.53 Å. Hence, the Pd/WSe₂ interlayer barrier height is significantly less than that of WSe₂/EG. Thus, the barrier arising from the interlayer gap at the WSe₂/EG interface dominates the electronic transport. It is to be noted that our calculations also show the existence of the conventional Schottky barrier of 0.95 eV between the Pd electrode and the monolayer WSe₂ [35].

5.4   Conclusions

In this chapter, we demonstrate a synthetic route to forming WSe_2/EG heterostructures via vdW epitaxy. Even though the lattice mismatch between WSe_2 and graphene is shown to be 23%, the heterostructure is commensurate at every third W and fourth C atom, indicating the potential to grow single crystal heterostructures over large areas. Additionally, we provide evidence that the structural, chemical, and optical properties of the WSe_2 grown on graphene match or exceed that of mechanically exfoliated WSe_2 films. Finally, WSe_2/EG diode structures and C-AFM indicate that efficient tunneling is possible through the WSe_2 layer to graphene, and the primary source of tunneling resistance occurs at the interlayer gap between the WSe_2 and graphene layer. The next chapter discusses an optimum vertical electrical transport achieved by modulating the Fermi level of the EG substrate.

References

1. Geim, A.K., Grigorieva, I.V.: Van der Waals heterostructures. Nature. **499**, 419–425 (2013)
2. Britnell, L., et al.: Field-effect tunneling transistor based on vertical graphene heterostructures. Science. **335**, 947–950 (2012)
3. Haigh, S.J., et al.: Cross-sectional imaging of individual layers and buried interfaces of graphene-based heterostructures and superlattices. Nat. Mater. **11**, 764–767 (2012)
4. Lin, Y.-C., et al.: Direct synthesis of van der Waals solids. ACS Nano. **8**, 3715–3723 (2014)
5. Liu, Z., et al.: Direct growth of graphene/hexagonal boron nitride stacked layers. Nano Lett. **11**, 2032–2037 (2011)
6. Shi, Y., et al.: Van der Waals epitaxy of MoS_2 layers using graphene as growth templates. Nano Lett. **12**, 2784–2791 (2012)
7. Levendorf, M.P., et al.: Graphene and boron nitride lateral heterostructures for atomically thin circuitry. Nature. **488**, 627–632 (2012)
8. Ohta, T., et al.: Evidence for interlayer coupling and Moiré periodic potentials in twisted bilayer graphene. Phys. Rev. Lett. **109**, 186807 (2012)
9. Robinson, J.T., et al.: Electronic hybridization of large-area stacked graphene films. ACS Nano. **7**, 637–644 (2013)
10. Wallace, R.M.: In-situ characterization of 2D materials for beyond CMOS applications. ECS Trans. **64**, 109–116 (2014)
11. Herrera-Gómez, A., Hegedus, A., Meissner, P.L.: Chemical depth profile of ultrathin nitrided SiO_2 films. Appl. Phys. Lett. **81**, 1014 (2002)
12. Lin, Y.-C., et al.: Atomically thin heterostructures based on single-layer tungsten diselenide and graphene. Nano Lett. **14**, 6936–6941 (2014)
13. Robinson, J.A., et al.: Contacting graphene. Appl. Phys. Lett. **98**, 053103 (2011)
14. Das, S., Appenzeller, J.: Where does the current flow in two-dimensional layered systems? Nano Lett. **13**, 3396–3402 (2013)
15. Fang, H., et al.: High-performance single layered WSe_2 p-FETs with chemically doped contacts. Nano Lett. **12**, 3788–3792 (2012)
16. de Heer, W.A., et al.: Epitaxial graphene. Solid State Commun. **143**, 92–100 (2007)
17. Robinson, J., et al.: Nucleation of epitaxial graphene on SiC(0001). ACS Nano. **4**, 153–158 (2010)

18. Yeh, P.-C., et al.: Probing substrate-dependent long-range surface structure of single-layer and multilayer MoS_2 by low-energy electron microscopy and microprobe. Phys. Rev. B. **89**, 155408 (2014)

19. Zhang, Y., et al.: Direct observation of the transition from indirect to direct bandgap in atomically thin epitaxial $MoSe_2$. Nat. Nanotechnol. **9**, 111–115 (2014)

20. Ohta, T., Beechem, T.E., Robinson, J.T., Kellogg, G.L.: Long-range atomic ordering and variable interlayer interactions in two overlapping graphene lattices with stacking misorientations. Phys. Rev. B. **85**, 075415 (2012)

21. Wilson, J.A., Yoffe, A.D.: The transition metal dichalcogenides discussion and interpretation of the observed optical, electrical and structural properties. Adv. Phys. **18**, 193–335 (1969)

22. Koma, A.: Van der Waals epitaxy—a new epitaxial growth method for a highly lattice-mismatched system. Thin Solid Films. **216**, 72–76 (1992)

23. Ji, Q., et al.: Epitaxial monolayer MoS_2 on mica with novel photoluminescence. Nano Lett. **13**, 3870–3877 (2013)

24. Hibino, H., et al.: Microscopic thickness determination of thin graphite films formed on SiC from quantized oscillation in reflectivity of low-energy electrons. Phys. Rev. B. **77**, 075413 (2008)

25. Riedl, C., Coletti, C., Iwasaki, T., Zakharov, A.A., Starke, U.: Quasi-free-standing epitaxial graphene on SiC obtained by hydrogen intercalation. Phys. Rev. Lett. **103**, 246804 (2009)

26. Zhang, C., Johnson, A., Hsu, C.-L., Li, L.-J., Shih, C.-K.: Direct imaging of band profile in single layer MoS_2 on graphite: quasiparticle energy gap, metallic edge states, and edge band bending. Nano Lett. **14**, 2443–2447 (2014)

27. Weng, X., et al.: Structure of few-layer epitaxial graphene on 6H-SiC(0001) at atomic resolution. Appl. Phys. Lett. **97**, 201905 (2010)

28. McDonnell, S., et al.: Hole contacts on transition metal Dichalcogenides: Interface chemistry and band alignments. ACS Nano. **8**, 6265–6272 (2014)

29. Huang, J.-K., et al.: Large-area synthesis of highly crystalline WSe_2 monolayers and device applications. ACS Nano. **8**, 923–930 (2014)

30. Zhang, W., et al.: Ultrahigh-gain photodetectors based on atomically thin graphene-MoS_2 heterostructures. Sci. Rep. **4**, 3826 (2014)

31. Shim, G.W., et al.: Large area single layer $MoSe_2$ and its van der Waals Heterostructures. ACS Nano. **8**, 6655–6662 (2014)

32. Mouri, S., Miyauchi, Y., Matsuda, K.: Tunable photoluminescence of monolayer MoS_2 via chemical doping. Nano Lett. **13**, 5944–5948 (2013)

33. Buscema, M., Steele, G.A., van der Zant, H.S.J., Castellanos-Gomez, A.: The effect of the substrate on the Raman and photoluminescence emission of single-layer MoS_2. Nano Res. **7**, 561–571 (2014)

34. Coy Diaz, H., Addou, R., Batzill, M.: Interface properties of CVD grown graphene transferred onto MoS_2 (0001). Nanoscale. **6**, 1071–1078 (2014)

35. Das, S., Appenzeller, J.: WSe_2 field effect transistors with enhanced ambipolar characteristics. Appl. Phys. Lett. **103**, 103501 (2013)

Chapter 6
Tuning Electronic Transport in WSe$_2$-Graphene

6.1 Introduction

In many atomically thin photovoltaic devices, field effect transistors, and tunneling diodes, 2D TMDC have been used as a semiconducting layer in tandem with graphene and many other substrates. It is necessary to achieve efficient charge transport across WSe$_2$-graphene, which creates a semiconductor to semimetal junction. In such cases, the band alignment engineering is required to ensure a low-resistance, ohmic contact. In previous chapter, we cover preparation and fundamental properties of WSe$_2$-graphene. In this chapter, we investigate the impact of graphene properties on the transport at the interface of WSe$_2$-graphene. Electrical transport measurements reveal a change in resistance between WSe$_2$ and fully hydrogenated epitaxial graphene (EG$_{FH}$) compared to WSe$_2$ grown on partially hydrogenated epitaxial graphene (EG$_{PH}$). Using low-energy electron microscopy and reflectivity (LEEM/LEER) on these samples, we extract the work function difference between the WSe$_2$ and graphene and employ a charge transfer model to determine the WSe$_2$ carrier density in both cases. The results here indicate that WSe$_2$-EG$_{FH}$ displays nearly ohmic behavior at small biases due to a large hole density in the WSe$_2$, whereas WSe$_2$-EG$_{PH}$ forms a Schottky barrier junction.

Monolayer WSe$_2$ could be used as a tunneling barrier in the WSe$_2$-EG diodes [1, 2]. However, the turn-on voltages (1.5–1.8 V) in the I–V characteristics of this vertical diode are considered large and tentatively attributed to the unfavorable n-type nature of the EG [1]. In this chapter, we investigate the role of the carrier type of the graphene and show that it is a critical parameter in controlling the charge transport at the WSe$_2$/graphene or with other 2D semiconductor interface. Traditional mechanical transfer processes also utilized transferred graphene, which is inherently p-type due to water and environmental doping. These dopants ultimately control the electrical characteristics of the heterostructure stack. By controlling the doping type and concentration of EG from n- to p- *via* in situ hydrogen intercalation

© Springer Nature Switzerland AG 2018
Y.-C. Lin, *Properties of Synthetic Two-Dimensional Materials and Heterostructures*, Springer Theses,
https://doi.org/10.1007/978-3-030-00332-6_6

during the WSe$_2$ synthesis [3], we demonstrate the origins of ohmic behavior in TMDC/graphene structures and obtain a reduced resistance ohmic transport between a WSe$_2$-graphene interface. Low-energy electron microscopy (LEEM), low-energy electron reflectivity (LEER), and conductive atomic force microscopy (CAFM) were performed on monolayer WSe$_2$-EG$_{PH}$ and monolayer WSe$_2$-EG$_{FH}$, with the results showing that use of EG$_{FH}$ (p-type graphene) as the bottom electrode of WSe$_2$-graphene diodes can lead to a reduced resistance of *I–V* characteristics.

6.2 Experimental Methods

Epitaxial graphene was synthesized via silicon sublimation of the silicon face of 6H silicon carbide [6H SiC (0001)] at 1625 °C in a 200 Torr Ar environment inside a heating chamber made of pure graphite. The SiC substrates were pre-etched at 700 Torr with flowing 10% H$_2$-Ar mixtures (total flow 500 sccm) to remove subsurface damage due to substrate polishing [4]. The WSe$_2$-EG growth is graphically illustrated in Fig. 6.1a [2]. The precursors chosen for WSe$_2$ synthesis are tungsten hexacarbonyl (W(CO)$_6$) and dimethylselenium ((CH$_3$)$_2$Se), which provide the W and Se, respectively. In order to eliminate carbon contamination from the precursor, a 100% H$_2$ environment is utilized for WSe$_2$ synthesis. This necessity of 100% H$_2$ significantly modulates the chemical environment of graphene, comparing to a dilute H$_2$ environment for WSe$_2$ growth. X-ray photoemission spectroscopy (XPS), equipped with a monochromatic Al-Kα source (E = 1486.7 eV) and an Omicron Argus detector operating with pass energy of 15 eV, carried out on the samples grown via powder vaporization (PV) using 5% H$_2$ at 900 °C and MOCVD using 100% H$_2$ at 800 °C, both in the same growth time, shows that the C1s core level of the WSe$_2$-EG via MOCVD has been shifted toward a lower binding energy by 0.4 eV compared to that of WSe$_2$-EG grown via PV [8]. This shift of C1s core level

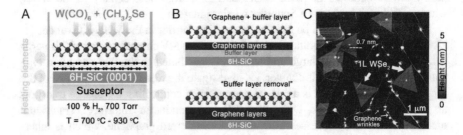

Fig. 6.1 (**a**) Illustration of MOCVD process of WSe$_2$ monolayer on EG-SiC and the synthesis conditions. (**b**) When the process of WSe$_2$ synthesis is carried out at a lower temperature, the buffer layer decoupling is incomplete (top). A higher synthesis temperature can efficiently convert the buffer layer into a layer of graphene via hydrogen intercalation (bottom). (**c**) AFM image of WSe$_2$-EG heterostructure. Monolayers are mostly 0.7 nm in height. The wrinkles of graphene can be seen, which promoted vertical WSe$_2$ growth [8]

in EG has been associated with hydrogen intercalation [3, 5]. Evident from XPS, the 100% H_2 environment leads to complete hydrogen interaction at the EG/SiC interface, fully passivating the buffer at 900 °C.

The growth of WSe_2 on EG proceeds by vdW epitaxy, mediating the high lattice mismatch between WSe_2 and graphene [6]. Tungsten selenide (WSe_2) atomic layers are grown via MOCVD on EG/SiC substrates employing H_2 as a carrier gas [2] at 800 °C and 930 °C in order to study how hydrogen intercalation impacts the electrical transport between graphene and WSe_2, for the different growth temperatures (Fig. 6.1b). In order to eliminate carbon contamination in the WSe_2 [2], a 100% H_2 environment is utilized. After 30 min for growth, the as-grown atomic layers were confirmed to be monolayer WSe_2, 1 μm wide and 0.7 nm thick with atomic force microscopy (AFM) (Fig. 6.1c). The optical bandgap of monolayer WSe_2 measured via photoluminescence (PL) spectroscopy is found to be 1.63 eV (Fig. 6.2a), using a fitted Lorentzian peak function [1, 2]. The Raman spectra exhibit two peaks of WSe_2 at 251 cm^{-1} and 260 cm^{-1} assigned to $E^1_{2g} + A_{1g}$ and 2LA, respectively (Inset, Fig. 6.2a) [2]. The B^1_{2g} peak at 310 cm^{-1} typically seen in bi- and multilayer WSe_2 is absent, verifying the presence of monolayer [2, 7]. Raman and PL measurements were performed in a WITec confocal Raman microscope with a 488 nm wavelength laser. The topographic AFM micrographs were taken in a Bruker Dimension with a scan rate of 1 Hz. The current-to-voltage (I–V) measurements performed on the samples were done in the same AFM with a PtIr-coated tip. Low-energy electron microscopy (LEEM) characterization was performed with an Elmitech III low-energy electron microscope. From sequences of images acquired with 0.1 eV energy spacing, Low-energy electron reflectivity (LEER) spectra were extracted at specific spatial locations on the surface.

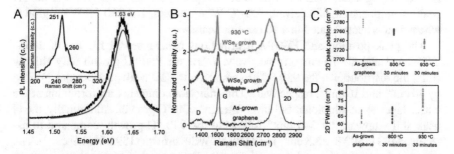

Fig. 6.2 (a) PL peak at 1.63 eV corresponding to the optical bandgap of monolayer WSe_2. Inset is the Raman peaks of the same spot as PL. Peaks at 251 cm^{-1} and 260 cm^{-1} are assigned to $E^1_{2g}(\Gamma)$/$A_{1g}(\Gamma)$ and $A_{1g}(M)$/2LA(M) of WSe_2 crystals, respectively [2]. (b) Higher growth temperature of WSe_2 can enhance the in situ hydrogenation on EG, evident by comparing Raman spectra of EG from 800 °C and 930 °C process. As-grown EG is present as reference. All of the spectrums were deconvoluted with SiC background. (c) and (d) Among the Raman 2D peak positions and corresponding FWHM from as-grown, 800 °C process and 930 °C process, the 930 °C process resulted in the largest position shift as well as width broadening, indicating increased graphene thickness and strain release due to the decoupling of buffer layers [8]

6.3   Results and Discussion

6.3.1   WSe$_2$ Synthesis and Buffer-Layer Decoupling

WSe$_2$ monolayer can be synthesized over a range of temperatures. However, the electrical properties of EG can be modified significantly at higher growth temperatures [5]. As-received EG confines a C-rich "buffer layer" at the EC/SiC interface on a single-crystalline SiC substrate. The C-rich buffer layer can be converted to a new layer of graphene by passivating the EC/SiC interface with hydrogen atoms [3, 5]. Furthermore, it has been demonstrated that the transformation of the buffer into graphene is more efficient at a higher temperature during the hydrogenation [5]. The Raman spectra of EG (D, G, and 2D peaks) [9] before and after WSe$_2$ growth at 800 °C and 930 °C were compared to understand the impact of WSe$_2$ growth temperatures on structural properties of the EG. In the case of the 800 °C, 30 min growth of WSe$_2$, the D-peak intensity of EG increases as evidenced by a higher D/G ratio. This can be attributed to partial hydrogenation of the epitaxial graphene (EG$_{PH}$) that only converts a small portion of the buffer into graphenes [5, 9]. On the other hand, in the case of the 930 °C (also 30 min WSe$_2$ growth), the D/G ratio is notably smaller. This evolution of the D peak as the hydrogenation is increased indicates a more complete buffer conversion [5].

The Raman 2D peak of EG has been a reliable indicator for the information on graphene, such a thickness, layer stacking, and a measure of compressive strain induced by graphene/SiC lattice mismatch [10]. In this work, the 2D peak of the 930 °C growth is at a lower peak position, with broader full width half maximum (FWHM), compared to that of the growth at 800 °C. This is associated with a combination of increased layer thickness and released compressive strain [5]. This trend is evident (Fig. 6.2c, d) as an evolution of the peak position and FWHM of the 2D peak of graphene in as-received, 800 °C WSe$_2$ growth, and 930 °C WSe$_2$ growth where data is accumulated in a 10 μm-line scan.

The peak position and FWHM distributions of as-received EG are of 2765–2782 cm^{-1} and 60–70 cm^{-1}, respectively. While the 800 °C growth only slightly shifts the distributions, the 930 °C growth shifts the 2D peak position and FWHM by 40 cm^{-1} and 10 cm^{-1}, respectively. This indicates complete transformation of the buffer layer to an additional EG layer after the 930 °C growth. Additionally, the G peak shifts upward by 3–5 cm^{-1} following the 930 °C growth, compared to as-received EG (1592.8 ± 3.5 cm^{-1}) and 800 °C WSe$_2$ growth (1593.5 ± 1.9 cm^{-1}). The wavenumber of the G peak increases when the Fermi level of graphene moves away from the Dirac point toward both n- and p-type direction [11]. It has been reported that the buffer conversion shifts the Fermi level to a point below the Dirac point [12]. Therefore, the stiffening G peak of the EG from the 930 °C growth is contributed to a conversion from n-type to p-type graphene due to the buffer conversion, as consistent with the evolution of the 2D peak (Fig. 6.2c, d). Despite the G peak is a good indicator on the doping effects on graphene, the 2D peak is chosen since its peak position has a larger shift than G peak in response to hydrogenation/buffer conversion.

6.3.2 LEEM/LEER Measurements and Analysis

In order to study the surface and electronic structure of theWSe$_2$-EG heterostructure, low-energy electron microscopy (LEEM) is employed. In addition, low-energy electron reflectivity (LEER) spectra provide an accurate means of determining the number of graphene layers as well as extracting the work function differences over the sample surface [13, 14]. The LEEM images of WSe$_2$-EG from the 800 C growth show triangular WSe$_2$ with an edge length of 1 µm; it also shows that the nucleation preferentially occurs near SiC step edge (Fig. 6.3a). The graphene is found predominantly in monolayer plus buffer form, but small area of bi- and trilayer graphene is also found. This confirms that the buffer is most likely intact as as-received EG, although a negligible portion of it could be eliminated during the 800 °C growth.

Low-energy electron reflectivity spectra (LEER) show oscillations for graphene and WSe$_2$ for the respective regions of the surface and also allow material identification in the LEEM images, as shown in Fig. 6.3b for the 800 °C growth [15]. In addition, LEER curves permit determination of the local work function on the surface,

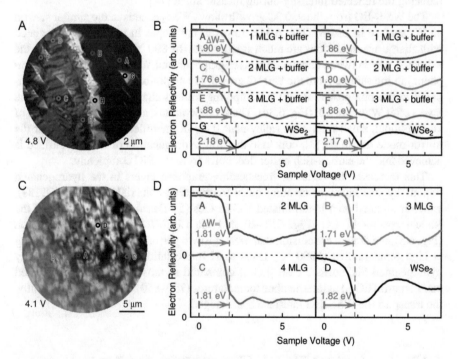

Fig. 6.3 (a) LEEM image of WSe$_2$ grown on EG-SiC at 800 °C (EG$_{PH}$), acquired at sample voltage of 6.2 V. Labeled points indicate location of reflectivity spectra in (b), which are used to identify the materials in the image. Bright triangles are WSe$_2$ islands; dark regions are mono- to multilayer graphene on carbon-rich buffer layer. Δǔ4A; value, to the left of each spectrum in (b), quantifies the electrostatic potential surface variation and hence the variation of the vacuum level. (c) LEEM image of WSe$_2$ grown on EG-SiC at 930 °C (EG$_{FH}$). (d) Reflectivity spectra of the points labeled in the (c); characteristic of a released buffer layer (due to passivated SiC dangling bonds). ΔW values show smaller variation than in the WSe$_2$-E$_{PH}$ case [8]

since for sufficiently low sample voltages (~2 V), the incident electrons are totally reflected from the sample (i.e., the "mirror-mode" transition). This voltage of this mirror-mode transition corresponds to the work function difference (ΔW) between the surface of EG, WSe$_2$, and WSe$_2$-EG and the LEEM electron emitter. Fitting of these transition voltages (or energies) permits the extraction of the ΔW across the surface [2]. A difference of $\Delta W \approx 0.31 \pm 0.03$ eV is found between the work functions of graphene and WSe$_2$ (uncertainty comes from a few factors, like uncertainties in the measurement, analysis, and variations of the sample surface). This observed ΔW is between "WSe$_2$ in contact with underlying graphene" (G in Fig. 6.3a) and "a bare graphene region without WSe$_2$" (A or B in the Fig. 6.3a). The presence of interface dipoles and a change in local work function implies charge transfer between the WSe$_2$ and the graphene below. Consistent with this interpretation, it is noted that LEER curves measured on the WSe$_2$ from the 800 °C growth (Fig. 6.3b) display a broad, sloping feature for voltages below the mirror-mode transition. This feature also indicates the presence of electric dipoles on the edges of the triangular WSe$_2$ which displace the incident and reflected electron beam, thus reducing the reflected intensity during measurement [14].

The WSe$_2$-EG from the 930 °C growth shows WSe$_2$ islands in the similar size on an EG surface in LEEM (Fig. 6.3c); however, the slopes in reflectivity associated with charge accumulations are much smaller than the 800 °C sample. Moreover, the extracted ΔW between uncovered bilayer graphene and WSe$_2$ (in contact with graphene) in the sample from the 930 °C growth are negligible (0.03 ± 0.03 eV) compared to the one grown at 800 °C, implying limited charge transfer between the layers after WSe$_2$ growth (Fig. 6.3d). These observations, along with the presence of an additional, small minimum valley near 0 eV in the reflectivity spectra near the mirror-mode transition [3], conclude that full hydrogenation of the SiC surface is achieved and the carbon-rich buffer decouples from the SiC completely.

This increases the count of freestanding graphene layers in the hydrogenated regions by one layer, or creating quasi-freestanding-epitaxial graphene (QFEG), which is situated on a H-terminated SiC surface [5]. Based on the evolution of the Raman spectra of EG (Fig. 6.2b–d) and the LEEM/LEER data interpretation (Fig. 6.3), we draw a conclusion that WSe$_2$ growth performed in a MOCVD setting at high temperatures (>900 °C) leads to H intercalation and formation of fully hydrogenated SiC surface case (EG$_{FH}$) compared to those partially hydrogenated counterparts (EG$_{PH}$) at intermediate temperatures (750–850 °C) [3, 5]. Concurrently, the transport the WSe$_2$/EG interface has changed.

6.3.3 Conductive AFM I–V Characteristics and Band Alignment Model

EG residing on top of the buffer layer on 6H-SiC (0001) is n-type doped [4, 5, 16] due to the combination of bulk and interface donor states [12, 17] and has a Fermi energy 0.45 eV above the Dirac point [12]. In contrast, QFEG is known to be p-type

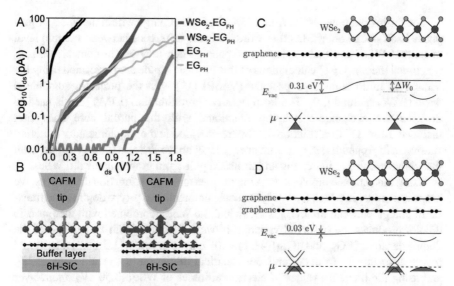

Fig. 6.4 Electrical measurements (**a**) show the *I–V* curves of EG$_{PH}$, EG$_{FH}$, WSe$_2$-EG$_{PH}$, and WSe$_2$-EG$_{FH}$. WSe$_2$-EG$_{PH}$ is more resistive than that of WSe$_2$-EG$_{FH}$, indicating the barrier to transport is larger for WSe$_2$-EG$_{PH}$. (**b**) The WSe$_2$-EG$_{PH}$ resulted in a small tunneling current (left), while the tunneling current is magnified after the decoupling of buffer layer (right). The yellow surface of CAFM tips symbols PtIr coatings. Band alignment and vacuum energy diagrams for the two heterostructures, WSe$_2$-EG$_{PH}$ (**c**) and WSe$_2$-EG$_{FH}$ (**d**), showing variations of vacuum energy of the surface due to partial WSe$_2$ coverage [8]

doped [3, 12]. The hydrogenation process changes the electrical properties of graphene on SiC significantly. This change is due to the presence of the spontaneous polarization of the hexagonal 6H-SiC, which lowers the Fermi level to a position below the Dirac point by near 0.30 eV once the hydrogenation is completed [12, 18]. This modification in the doping of graphene influences the transport across the WSe$_2$-graphene interface. In order to elucidate the transport properties, vertical *I–V* measurements were performed on both of the 800 °C and 930 °C samples (labeled as WSe$_2$-EG$_{PH}$ and WSe$_2$-EG$_{FH}$, respectively) in conductive AFM (CAFM). A CAFM tip with PtIr coating and the underneath graphene are as source and drain, respectively. The WSe$_2$-EG$_{FH}$ diode has increasing current that starts at near zero bias (Fig. 6.4a). The main component of the current near zero bias for WSe$_2$-EG$_{PH}$ is due to tunneling from the CAFM tip to graphene through the WSe$_2$ gap. On the other hand, for WSe$_2$-EG$_{FH}$, the WSe$_2$ layer seems to act as a short between the CAFM tip and the EG$_{FH}$ (Fig. 6.4b).

Our LEEM measurements and analysis above indicate a ΔW of 0.31 eV between the WSe$_2$ (in contact with EG$_{PH}$) and the uncovered monolayer EG$_{PH}$, while the ΔW between the WSe$_2$ (in contact with EG$_{FH}$) and the uncovered bilayer EG$_{FH}$ is near zero. The measured ΔW is a combination effect of intrinsic interface dipole and extrinsic interface dipole. The extrinsic dipole is due to doping caused by charge transfer between WSe$_2$ and graphene.

The intrinsic dipole results from redistribution of charge within the WSe_2 or graphene itself upon contact. In other words, it is the difference between vacuum level of neutral WSe_2 and that of neutral graphene when they are in contact. Density functional theory (DFT) calculations of this intrinsic dipole are performed using the Vienna ab initio simulation package (VASP) [19] with the projector-augmented wave (PAW) method [20]. The local density approximation (LDA) [21] is used to describe the exchange-correlation functional with the partial core correction included. More DFT calculation details are provided in the supplementary material. As shown in Appendix, the vacuum energy level above WSe_2 is 0.17 eV higher than that above graphene, indicating an intrinsic dipole from graphene toward WSe_2.

Using this intrinsic dipole, along with the measured work function differences, we propose a model in which the WSe_2 has some unintentional p-type doping, and transfer of charge between the EG_{PH} or EG_{FH} and the WSe_2 (combined with the intrinsic dipole) produces the observed variation in work function. With knowledge of the doping density of EG_{PH} and EG_{FH} [$(4 \pm 1) \times 10^{12}$ cm^{-2} n-type and $(1.5 \pm 0.2) \times 10^{13}$ cm^{-2} p-type, respectively, from our previous electrical studies on EG_{PH} and EG_{FH}] [5, 22], and using the literature values of electron affinities of WSe_2 (3.50 eV), monolayer graphene (4.57 eV), and bilayer graphene (4.71 eV) [23], we then compute the transfer of charge between the WSe_2 and the EG_{PH} or EG_{FH}. This charge transfer, for a given, unintentional doping density of the WSe_2, yields theoretical values for the ΔW; the doping density is determined by matching these differences to experiment. Our model is illustrated in Fig. 6.4c, d. (The dependence of the results on the electron affinities of WSe_2 and graphene is discussed in Appendix B.)

For the charge transfer computation, we employ the standard linear band structure around the K point for the monolayer graphene from EG_{PH} and hyperbolic bands near the band extrema for bilayer graphene from EG_{FH} and for WSe_2 around K points, based on tight-binding models [24, 25]. The method to compute the electrostatics is similar to that described by Li et al. [26] in Fig. 6.4c, d showing band diagrams of the WSe_2-EG_{PH} and WSe_2-EG_{FH} at their interfaces, which are graphene partially covered by WSe_2. Both the intrinsic interface dipole and the charge transfer are taken into consideration, and equilibrium is reached when the Fermi levels are aligned. The difference between the vacuum energy of WSe_2 (contacting graphene) and the underlying graphene (e.g., ΔW_0 in Fig. 6.4c) is thus a combined effect of the intrinsic interface dipole and the charge transfer.

In order to match the theoretical work function difference between the WSe_2 (in contact) with the uncovered graphene with the experimental values (0.31 eV and 0.03 eV for WSe_2-EG_{PH} and WSe_2-EG_{FH}, respectively), we employ an unintentional p-type doping of 1.3×10^{12} cm^2 for the WSe_2 before charge transfer between the WSe_2 and the underlying graphene. When the WSe_2 is put in contact with EG_{PH} (n-type), electrons transfer from the EG_{PH} to the WSe_2, leading to nearly complete compensation of the p-type doping in the WSe_2 and a negligible carrier density in the WSe_2. The Fermi level ends up well inside the bandgap of the WSe_2 and near the charge neutrality point in the graphene (Fig. 6.4c). For the case of the WSe_2 in contact with the EG_{FH} (which is p-type), the same unintentional doping of the WSe_2 is employed (1.4×10^{12} cm^{-2}, p-type). To reach equilibrium, electrons transfer from the WSe_2 to the EG_{FH}, making the WSe_2 more p-type (carrier density $2.9 \times 10^{12} \, cm^{-2}$). The predicted Fermi level of the WSe_2-EG_{FH} remains near the top of the valence band of its WSe_2. From an electrical point of view, the WSe_2 on the EG_{PH} forms a Schottky barrier (i.e., leading to a low conductivity), whereas the WSe_2 on the EG_{FH} forms a better contact (i.e., higher conductivity), leading to a $\sim 10^3 \times$ increase in current in magnitude (Fig. 6.4a). An additional output of our charge transfer computations is the sum of the bandgap plus electron affinity of the WSe_2, $\chi_{WSe2} + E_g$ (only the sum enters, since the electron density in the WSe_2 conduction band is negligible). In order to match the observed work function variations, we deduce an unintentional doping density in the WSe_2 of $1.3 \times 10^{12} \, cm^{-2}$, and the value of $\chi_{WSe2} + E_g$ is determined to be 5.1 eV. This value is consistent with an electron affinity of ~ 3.1 eV for WSe_2 using first-principles GW calculation from the literature [27], together with a bandgap of ~ 2 eV, which is in agreement with several recently reported experimental values [28, 29].

6.4 Conclusions

By varying the temperatures for growth of WSe_2 on as-received EG in a H_2-rich MOCVD process, the electrical transport across WSe_2-graphene is controllable. This investigation combining LEED/LEEM, Raman spectra, and electrical measurements confirmed the transport across the WSe_2-graphene interface is controlled by the doping of EG on SiC, which is tuned by the presence of a carbon-rich buffer layer. The band alignment diagrams of two different heterostructures were constructed with the measured ΔW between the WSe_2 and the graphene from LEER. Taking into account their intrinsic interface dipoles and charge transfer, the diagrams indicate the Schottky barrier presenting in WSe_2-EG_{PH} and a reduced barrier in WSe_2-EG_{FH}, which are in agreement with their I–V characteristics. The work demonstrates that doping of the graphene plays a role in the transport in these atomically thin heterointerfaces. Epitaxial graphene on SiC is unique because it can be made of n- or p-type based on the TMDC growth conditions. This allows for one to engineer the transport across a truly pristine 2D material interface.

References

1. Lin, Y.-C., et al.: Atomically thin heterostructures based on single-layer tungsten diselenide and graphene. Nano Lett. **14**, 6936–6941 (2014)
2. Eichfeld, S.M., et al.: Highly scalable, atomically thin WSe_2 grown via metal-organic chemical vapor deposition. ACS Nano. **9**, 2080–2087 (2015)
3. Riedl, C., Coletti, C., Iwasaki, T., Zakharov, A.A., Starke, U.: Quasi-free-standing epitaxial graphene on SiC obtained by hydrogen intercalation. Phys. Rev. Lett. **103**, 246804 (2009)
4. Emtsev, K.V., et al.: Towards wafer-size graphene layers by atmospheric pressure graphitization of silicon carbide. Nat. Mater. **8**, 203–207 (2009)

5. Robinson, J.A., et al.: Epitaxial graphene transistors: enhancing performance via hydrogen intercalation. Nano Lett. **11**, 3875–3880 (2011)
6. Koma, A.: Van der Waals epitaxy—a new epitaxial growth method for a highly lattice-mismatched system. Thin Solid Films. **216**, 72–76 (1992)
7. Terrones, H., et al.: New first order Raman-active modes in few layered transition metal dichalcogenides. Sci. Rep. **4**, 4215 (2014)
8. Lin, Y.-C., et al.: Tuning electronic transport in epitaxial graphene-based van der Waals heterostructures. Nanoscale. **8**, 8947–8954 (2016)
9. Ferrari, A.C.: Raman spectroscopy of graphene and graphite: disorder, electron–phonon coupling, doping and nonadiabatic effects. Solid State Commun. **143**, 47–57 (2007)
10. Robinson, J.A., Puls, C.P., Staley, N.E., Stitt, J.P., Fanton, M.A.: Raman topography and strain uniformity of large-area epitaxial graphene. Nano Lett. **9**, 964–968 (2009)
11. Das, A., et al.: Monitoring dopants by Raman scattering in an electrochemically top-gated graphene transistor. Nat. Nanotechnol. **3**, 210–215 (2008)
12. Ristein, J., Mammadov, S., Seyller, T.: Origin of doping in quasi-free-standing graphene on silicon carbide. Phys. Rev. Lett. **108**, 246104 (2012)
13. Feenstra, R.M., et al.: Low-energy electron reflectivity from graphene. Phys. Rev. B. **87**, 041406 (2013)
14. Gopalan, D.P., et al.: Formation of hexagonal boron nitride on graphene-covered copper surfaces. J. Mater. Res. **31**, 945–958 (2016)
15. Vishwanath, S., et al.: Comprehensive structural and optical characterization of MBE grown MoSe$_2$ on graphite, CaF$_2$ and graphene. 2D Mater. **2**, 024007 (2015)
16. Ohta, T., et al.: Interlayer interaction and electronic screening in multilayer graphene investigated with angle-resolved photoemission spectroscopy. Phys. Rev. Lett. **98**, 206802 (2007)
17. Kopylov, S., Tzalenchuk, A., Kubatkin, S., Fal'ko, V.I.: Charge transfer between epitaxial graphene and silicon carbide. Appl. Phys. Lett. **97**, 112109 (2010)
18. Mammadov, S., et al.: Polarization doping of graphene on silicon carbide. 2D Mater. **1**, 035003 (2014)
19. Kresse, G., Furthmüller, J.: Efficient iterative schemes for ab initio total-energy calculations using a plane-wave basis set. Phys. Rev. B. **54**, 11169–11186 (1996)
20. Kresse, G., Joubert, D.: From ultrasoft pseudopotentials to the projector augmented-wave method. Phys. Rev. B. **59**, 1758–1775 (1999)
21. Ceperley, D.M., Alder, B.J.: Ground state of the electron gas by a stochastic method. Phys. Rev. Lett. **45**, 566–569 (1980)
22. Hollander, M.J., et al.: Heterogeneous integration of hexagonal boron nitride on bilayer quasi-free-standing epitaxial graphene and its impact on electrical transport properties. Phys. Status solidi. **210**, 1062–1070 (2013)
23. Yu, Y.-J., et al.: Tuning the graphene work function by electric field effect. Nano Lett. **9**, 3430–3434 (2009)
24. Liu, G.-B., Shan, W.-Y., Yao, Y., Yao, W., Xiao, D.: Three-band tight-binding model for monolayers of group-VIB transition metal dichalcogenides. Phys. Rev. B. **88**, 085433 (2013)
25. McCann, E., Koshino, M.: The electronic properties of bilayer graphene. Reports Prog. Phys. **76**, 056503 (2013)
26. (Oscar) Li, M., Esseni, D., Snider, G., Jena, D., Grace Xing, H.: Single particle transport in two-dimensional heterojunction interlayer tunneling field effect transistor. J. Appl. Phys. **115**, 074508 (2014)
27. Liang, Y., Huang, S., Soklaski, R., Yang, L.: Quasiparticle band-edge energy and band offsets of monolayer of molybdenum and tungsten chalcogenides. Appl. Phys. Lett. **103**, 042106 (2013)
28. He, K., et al.: Tightly bound excitons in monolayer WSe$_2$. Phys. Rev. Lett. **113**, 026803 (2014)
29. Zhang, C., et al.: Probing critical point energies of transition metal dichalcogenides: surprising indirect gap of single layer WSe$_2$. Nano Lett. **15**, 6494–6500 (2015)

Chapter 7
Atomically Thin Resonant Tunnel Diodes

7.1 Introduction

Vertical integration of 2D vdW materials is predicted to display novel electronic and optical properties absent in their constituent layers [1]. In this chapter the direct synthesis of two unique, atomically thin, multi-junction heterostructures is demonstrated by combining graphene with some important 2D TMDC: MoS_2, $MoSe_2$, and WSe_2, aiming to achieve "epitaxy-grade" material interfaces. Surprisingly, the realization of MoS_2-WSe_2-graphene and WSe_2-$MoSe_2$-graphene heterostructures leads to resonant tunneling in an atomically thin stack with spectrally narrow, room-temperature negative differential resistance characteristics.

Resonant tunneling of charge carriers occurs when spatially separated quantum states align; this leads to a unique current transport phenomenon known as negative differential resistance (NDR) [2, 3]. Novel nanoelectronic circuits that utilize bistability and positive feedback, such as novel memories, multivalued logic, inductor-free compact oscillators, and many other not-yet-realized electronic applications, would benefit from this feature [4, 5]. However, it is challenging to realize narrow NDR spectrum in a resonant tunneling diode (RTD) at room temperature due to carrier scattering related to interfacial imperfections, which are common in traditional semiconductor heterostructures made via epitaxy techniques [6]. Two-dimensional (2D) materials [7, 8] could ultimately address the interfacial imperfections that limit room-temperature NDR performance in traditional semiconductor interfaces to date with their vdW surfaces. While the direct growth techniques for multiple 2D materials are in its infancy [7], the majority of electronic transport and vertical integration in 2D materials has been reported using mechanically exfoliated flakes since 2004 [9]. There has been a collective effort to synthesize layered transition metal dichalcogenides (TMDC), with powder vaporization (PV) [10–12] paving the way for direct growth of atomically thin structures [10–15]. Moving forward from monolayered TMDC, vdW heterostructures, which are heterogeneous stacks of dissimilar

© Springer Nature Switzerland AG 2018

Y. -C. Lin, *Properties of Synthetic Two-Dimensional Materials and Heterostructures*, Springer Theses,
https://doi.org/10.1007/978-3-030-00332-6_7

atomic layers, have been believed to be the key for new properties not found in their constituent layers [16]. Although manual stacking has provided experimental verification of electronic bandgap modulations and strong interlayer coupling [17], it can also lead to contamination at material interfaces [18] that introduces unwanted scattering mechanisms and, therefore, inhibits the NDR. To overcome this obstacle, a synthetic route to achieve vdW heterostructures with cleaned interfaces will be critical to advance the field.

Here, we present the direct synthesis of MoS_2-WSe_2-graphene and WSe_2-$MoSe_2$-graphene heterostructures employing a combination of oxide PV and MOCVD. We not only demonstrate that these heterostructures exhibit the same interlayer electronic coupling found in mechanically exfoliated heterostructures [17, 19, 20] but also show that they exhibit unique electronic transport properties not typically found in exfoliated structures. We discover that direct grown heterostructures exhibit resonant tunneling of charge carriers, which leads to sharp NDR at room temperature. Importantly, we identify that the peak of the resonant tunneling can be tuned by modifying the stacking order or layer composition, which will be a powerful tool toward engineering heterostructures for ultralow-power electronic devices.

7.2   Experimental Methods

7.2.1   Epitaxial Graphene Grown on 6H-SiC

Graphene is synthesized on 6H-SiC (0001) in a graphite crucible [21]. The 6H-SiC substrate was annealed in H_2 at 1500 °C for 10 min in order to clean substrate surface prior to graphene growth. At this stage the chamber pressure is 700 Torr under a H_2 (50 s.c.c.m)/Ar (450 s.c.c.m) flow. After H_2 annealing the system temperature cooled to 850 °C and pumped/purged with ultra-high pure N_2 at least six times to remove H_2 gas. Subsequently the chamber is filled in Ar gas (500 s.c.c.m.) to 200 Torr. The chamber was then heated up to 1725 °C at 100 °C/min and dwelled at this temperature for 20 min to grow three layers of graphene within the terraces of substrates via sublimation of silicon on the silicon side of 6H-SiC (0001). The system cooled down naturally to room temperature after the growth.

7.2.2 MoS_2-WSe_2-EG and WSe_2-$MoSe_2$-EG Synthesis

WSe_2 can be grown on EG either via PV reaction of WO_3 and Se powders or via MOCVD [22, 23]. The PV reaction utilizes the vaporization of WO_3 powders in a ceramic boat placed at the center of 1″ horizontal hot wall tube reactor with a flow of H_2 (10 s.c.c.m.)/Ar (90 s.c.c.m.). The EG substrates for WSe_2 growth were placed at the downstream side of the tube and heated to 925 °C at 25 °C/min. Samples were held at 925 °C for 15 min and then cooled naturally to room temperature. The total pressure throughout the reaction is held at 6 Torr. Utilizing MOCVD, WSe_2 was synthesized in a vertical cold wall reactor using $W(CO)_6$ and DMSe precursors. The metallic organic precursors were transported into the reactor by carrier gas of 100% H_2 via a bubbler manifold that allows for controlling each precursor concentration independently. The Se to W ratio was fixed at 20,000. The MOCVD growth of WSe_2 took place at 800–850 °C with a total pressure of 700 Torr. The growth time varied between 15 and 30 min. After the completion of WSe_2-EG synthesis, an ex situ MoS_2 growth via the vapor phase reaction of MoO_3 and S powders was carried out in a horizontal hot wall tube reactor at 700 Torr. During the MoS_2 growth, MoS_3 powder in a ceramic boat placed at the center of heating zone was heated at 750 °C for 10 min. After the MoS_2 growth, the reactor cooled down to room temperature naturally. The processes for WSe_2-$MoSe_2$-EG synthesis are similar but steps reversed. The MoS_2 is grown first, followed by an ex situ WSe_2 growth via the vapor phase reaction of WO_3 and Se. A Se-S, ion exchange occurs in the MoS_2 converting the MoS_2 into $MoSe_2$ [24]. Subsequently, the WSe_2 layers grow on $MoSe_2$-EG as the hot zone is held at 950 °C for 45 min, resulting in WSe_2-$MoSe_2$-EG.

7.2.3 Materials Characterization

The samples are characterized using Raman spectroscopy, atomic force microscopy/conductive atomic force microscopy (AFM/CAFM), X-ray photoelectron spectroscopy (XPS), and transmission electron microscopy (TEM). A WITec CRM200 Confocal Raman microscope with 488/514/633 nm wavelength lasers was utilized for structural characterization. A Bruker Dimension with a scan rate of 0.5 Hz was utilized for the topography image during the AFM measurement. Conductive AFM (CAFM) measurement was performed in PeakForce TUNA mode with platinum (Pt) AFM tip. The applied voltage from tips to sample was increased from 0 to 2 V. The optimized loading force of the AFM tip and sensitivity was nominally 5 nN and 20 pA/V, respectively, for the I–V measurements carried out on the novel junctions. All the AFM/CAFM measurements in Bruker Dimension were at room temperature and in ambient. TEM cross-sectional samples were made via utilizing a NanoLab dual-beam FIB/SEM system. Protective layers of SiO_2 and Pt were deposited to protect the interesting region during focused ion beam milling. TEM imaging was performed using a JEOL 2100F operated at 200 kV.

For surface analysis, the sample was loaded into an ultra-high vacuum (UHV) with a base pressure lower than 2×10^{-10} mbar. The WSe_2/EG sample was then imaged using an Omicron variable temperature scanning tunneling microscope (STM) without any thermal treatment. The STM images were obtained at room temperature and in the constant current mode, with an etched tungsten tip. The same system is equipped with a monochromatic Al-Kα source (E = 1486.7 eV) and an Omicron Argus detector operating with pass energy of 15 eV. The spot size used during the acquisition is equal to 0.5 mm. Core-level spectra taken with 15 sweeps are analyzed with the spectral analysis software analyzer.

7.2.4 Theoretical Methods

In order to provide a theoretical explanation for the emerging phenomena presented, we perform quantum transport calculations by using density functional theory (DFT) coupled with the non-equilibrium Green's function (NEGF) formalism. Detailed setup for DFT and NEGF is placed in Appendix C [25].

7.3 Results and Discussion

7.3.1 Making Vertical vdW Heterostructures

The heterostructures are achieved by sequentially growing two dissimilar TMDC monolayers on multilayer (up to three layers) epitaxial graphene (EG) [21]. WSe_2 is synthesized using both routes: WO_3 and Se powders as the precursors for the PV route [22] and tungsten $W(CO)_6$ and DMSe for the MOCVD route [23]. MoS_2 layers are grown via vaporization of MoO_3 and S [11]. The heterostructure synthesis process is summarized in Fig. 7.1. The first TMD layer of the heterostructure, WSe_2 or MoS_2, is grown on tri-layer EG (Fig. 7.1a) at 950 °C and 750 °C for WSe_2-EG (Fig. 7.1b) and MoS_2-EG (Fig. 7.1c, d), respectively. Following the first step of TMDC growth, the surface coverage of the WSe_2 or MoS_2 on EG is above 60%, with averaged lateral size of 2 μm and 300 nm for WSe_2 and MoS_2. Next, a MoS_2-WSe_2-EG vertical heterostructure is created via a second growth of MoS_2 on WSe_2-EG at 750 °C in ex situ way (Fig. 7.1c). We find that wrinkles in the graphene as well as defects and edges within the WSe_2 promote vertical growth of the MoS_2 because those areas have a higher surface energy favoring island growth. On the other hand, bilayer of MoS_2/WSe_2 is primarily achieved in pristine regions of WSe_2 (Fig. 7.1c) [26].

The formation of the WSe_2-$MoSe_2$-EG heterojunction is somehow different to the growth of WSe_2 on MoS_2. During the synthesis, a selenium-sulfur ion exchange occurs when the MoS_2 is exposed to the selenium vapor just prior to the growth of WSe_2 that happens at 950 °C for 45 min [24]. Standard topographic characterization via AFM cannot clearly identify the location of the heterostructures (Fig. 7.1e, f); however, conductive AFM (CAFM) with *Pt-Ir* tip [28] provides a means to map the WSe_2-$MoSe_2$-EG junctions and adjacent WSe_2-EG regions due to a difference in heterostructure conductivity. Transmission electron microscopy (TEM) confirms the formation of crystalline, vertical vdW heterostructures (Fig. 7.2). Scanning transmission electron microscopy (STEM) (Fig. 7.2a, b) also verifies the heterostructure is not a manifestation of the alloying of the constituent TMDs, but indeed is a unique layer with pristine interfaces with atomic precision. In the case of MoS_2-WSe_2-EG, we have focused on a multilayer region of MoS_2-WSe_2 to ensure pristine layer formation beyond the ML. The clean interface between layers can be observed in both of STEM image and the EDS cross-profiles (Fig. 7.2c).

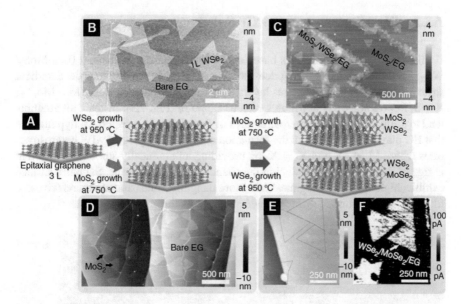

Fig. 7.1 (a) MoS_2-WSe_2-EG vertical heterostructures begin with the synthesis of 3L EG from SiC followed by (b) vapor transport or MOCVD of WSe_2 and (c) vapor transport of MoS_2. WSe_2-$MoSe_2$-EG heterostructures are similarly grown, except when (d) MoS_2 is grown first on EG followed by (e) growth of the WSe_2, a Se-S ion exchange occurs, leading to the formation of $MoSe_2$ from the original MoS_2 layer. The $MoSe_2$ domains are difficult to topographically identify; however, (f) conductive AFM clearly delineates their location due to enhanced tunneling at the heterostructures [27]

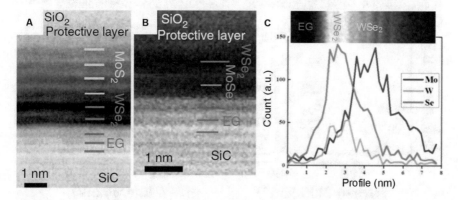

Fig. 7.2 Scanning TEM (a) and (b) confirms that the stacked structures exhibit pristine interfaces, with no intermixing of Mo-W or S-Se after synthesis. (c) The EDS cross-profiles in HAADF HRTEM of double-junction elemental distributions of MoS_2, WSe_2, and graphene and indicates no alloys in each TMDC layer [27]

7.3.2   2D Alloys on a Non-vdW Substrate

Unlike vertical heterostructures based on the same chalcogen element (i.e., MoS_2/WS_2) [29], the ordered layering does not occur when we attempt to grow a vertical structure based on heterogeneous layers where $M_1 \neq M_2$ and $X_1 \neq X_2$ (M = Mo, W; X = S, Se) on non-vdW substrates such as sapphire (Fig. 7.3). Instead, all attempts lead to alloying or lateral heterostructures of the layers. As a result, we hypothesize that EG plays a critical role in the formation of atomically precise vdW heterostructures where $M_1 \neq M_2$ and $X_1 \neq X_2$ by providing an atomically smooth surface that is free of dangling bonds, reducing the chance of ionic exchange. Sapphire surface exhibits higher surface roughness and more dangling bonds and are therefore more likely to impede surface diffusion, which catalyzes the alloying process.

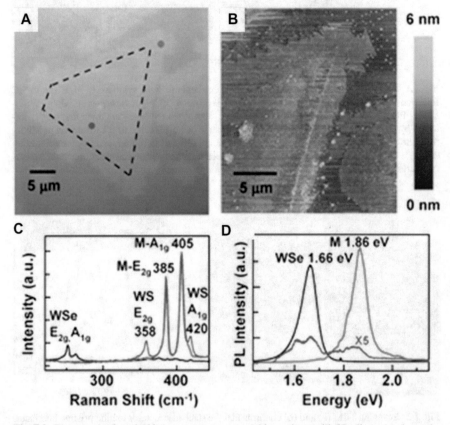

Fig. 7.3 The growth of vertical heterostructures on sapphire ends up with 2D alloys. (**a**) shows the optical micrograph after CVD process, and the boundary of per-growth WSe_2 part is located with black dashed line. In some case, we observed that MoS_2 is growth from the edge instead on the top of the WSe_2, which is clearly shown in (**a**). The AFM image of (**b**) confirms that the MoS_2 grows from the edge. Some structural damages on per-growth WSe_2 are found in (**b**). (**c**) and (**d**) show the Raman and PL spectrum of the WSe_2 before (black) and after (blue) and new-growth MoS_2 (red); the examined positions are indicated in (**a**) [27]

7.3.3 The Interlayer Coupling Within 2D Junctions

Most of the EG remains nearly defect free following the sequence of TMDC growths. However, there are regions of increased defectiveness due to either partial passivation of the graphene/SiC buffer layer [21] or formation of thick TMDC layers (Fig. 7.4a) [26]. Raman spectroscopy (Fig. 7.4b) also confirms the presence of significant fractions of ML WSe_2 (E_{2g}/A_{1g} at 250 cm^{-1} and 2LA at 263 cm^{-1}) [22] and MoS_2 (E_{2g} at 383 cm^{-1} and A_{1g} at 404 cm^{-1}) [26], as well as ML $MoSe_2$ (A_{1g} at 240 cm^{-1} and E^1_{2g} at 284 cm^{-1}) [24]. X-ray photoelectron spectroscopy [27] also corroborates the absence of any interaction between the two transition metal dichalcogenides or graphene; it also indicates that the MoS_2 exhibits an n-type nature, while the WSe_2 layer shows a p-type one. Semiconducting monolayer TMDCs exhibit a direct optical bandgap (E_{opt}) (MoS_2 at 1.8~1.9 eV, $MoSe_2$ at 1.55 eV, and WSe_2 at 1.6~1.65 eV) [30]. Their photoluminescence (PL) spectra provide evidence of electronic coupling between the layers (Fig. 7.4c). In addition to the standard PL peaks originating from the direct bandgap transition within the individual layers, the PL spectra of the heterostructures reveal interlayer excitons at 1.59 eV for MoS_2-WSe_2 and 1.36 eV for WSe_2-$MoSe_2$. Both of MoS_2-WSe_2 and WSe_2-$MoSe_2$ junctions exhibit type II band alignment [16, 17, 19, 31], where electrons in the WSe_2 conduction band transfer to the conduction band of MoS_2 ($MoSe_2$) and the excited holes in MoS_2 ($MoSe_2$) valence band transfer to the valence band of WSe_2, as described in Fig. 7.4d. Consistent with manually stacked heterojunctions in the literature [17, 19], the PL peak position is due to interlayer exciton recombination, confirming the carrier coupling at the heterojunction between the two monolayer TMDCs.

Additional evidence of coupling is provided by the topographical information of the heterostructures. Like graphene-hBN heterostructures [32], Moiré patterns of MoS_2-WSe_2 are observed in tapping-mode AFM (Fig. 7.5a), which are consistent with a small angular mismatch between MoS_2 and WSe_2. Furthermore, scanning tunneling microscopy/spectroscopy (STM/S) (Fig. 7.5b) confirms the presence of a Moiré pattern produced by the misorientation of MoS_2 relative to the WSe_2 layer at bottom. The lattice constant of the Moiré pattern is 9.8 ± 0.4 nm, which corresponds to a misorientation angle of ~1.9°. Modeling the heterostructure with this misorientation provides a consistent Moiré pattern, with a slightly smaller lattice constant of 9.6 nm (Inset, Fig. 7.5b). While the mechanical stacking technique leads to a variety of rotation angles between layers [17], the direct growth of vdW layers here appears to have a constrain on the rotational alignment, which may be critical for the optimization of coupling between the layers [33, 34].

Scanning tunneling spectroscopy further provides evidence that the hybridized bandgap of MoS_2-WSe_2 is significantly smaller than its WSe_2 counterpart (Fig. 7.5c). Based on STS, we infer that, for WSe_2-EG, the conduction band minimum (CBM) is at a sample bias of +0.71 ± 0.08 V and the valence band maximum (VBM) is at −1.11 ± 0.02 V. This indicates that the bandgap (E_g) of WSe_2 is 1.83 ± 0.05 eV, which is higher than E_{opt} (1.63 eV) due to the large excitonic binding energy in monolayer TMDCs [15, 20, 31, 35]. MoS_2-WSe_2-EG exhibits a CBM at +0.34 ± 0.03 V and VBM at −1.31 ± 0.03 V, indicating a hybridized, interlayer E_g

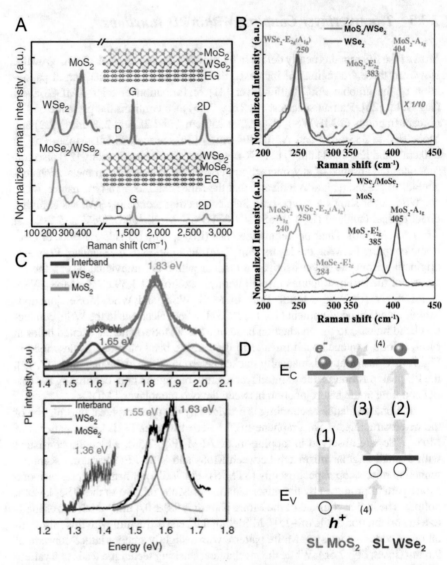

Fig. 7.4 (**a**) Raman spectra indicates preservation of the graphene has occurred during the synthesis process. (**b**) The spectrums clearly display distinct features from MoS_2/WSe_2 and $WSe_2/MoSe_2$ and indicate no alloy-like features. The asterisks indicate the signatures of their strong couplings and can also be found in mechanically stacked MoS_2/WSe_2. (**c**) The PL properties of MoS_2-WSe_2-EG and WSe_2-$MoSe_2$-EG reveal significant interlayer coupling, where the MoS_2-WSe_2-EG and WSe_2-$MoSe_2$-EG exhibit the intrinsic PL peaks corresponding to MoS_2, $MoSe_2$, and WSe_2 and also exhibit interband PL peaks at 1.59 and 1.36 eV. The charge transfer process of the bilayer systems is illustrated in (**d**), in which [1] and [2] denoted for intralayer excitons and [3] is for interlayer excitons [17]. The excitation wavelength (λ) is 488 nm and 633 nm for MoS_2-WSe_2 and WSe_2-$MoSe_2$, respectively [27]

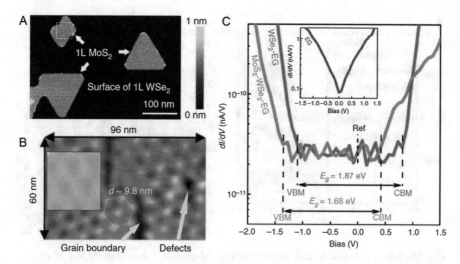

Fig. 7.5 (**a**) The Moiré patterns acquired via AFM in MoS_2 on WSe_2 indicate an alignment of nearly either 0° or 180° between the top and bottom layer due to their threefold symmetry, and (**b**) STM confirms the Moiré pattern with a lattice constant equal to (9.8 ± 0.4) nm. This structure can be reproduced theoretically when the misorientation angle between these layers is ~1.9° (inset). The continuity of the Moiré pattern is interrupted by the formation of a grain boundary and point defects, as indicated in the STM image. (**c**) STS on MoS_2-WSe_2-EG, WSe_2-EG, and EG (inset) provide evidence that the bandgap of the double-junction heterostructure (MoS_2-WSe_2-EG) is smaller than that of the single-junction (WSe_2-EG) heterostructure [27]

of 1.65 eV ± 0.02 V, which is slightly larger than its interlayer E_{opt} at 1.59 eV but smaller than the E_{opt} in monolayer MoS_2-EG [20, 31]. Mapping the tunneling current density of WSe_2-EG and WSe_2-$MoSe_2$-EG heterostructures via conductive AFM [28, 36] provides strong evidence that tunneling is much more readily achieved in WSe_2-$MoSe_2$-EG at a tip bias of 1.5 V, indicating a smaller, resonance tunneling, or both may be occurring. Finally, we note that defects, such as grain boundaries and vacancies, can disrupt the continuity of the Moiré pattern, emphasizing that imperfections in layers or at interfaces will impact the electronic behavior of vdW heterostructures.

7.3.4   Vertical Electrical Transport

Room-temperature current-voltage measurements through the heterostructures (carried out via CAFM) do not exhibit the traditional p-n junction diode-like transport found in mechanically exfoliated TMDC structures or direct grown single-junction (i.e., WSe_2-EG) structures [17, 26, 37]. Instead, we find that, following a "soft" turn-on, the current exhibits a peak at a certain bias voltage ($V_{peak} = +1.1$ V and +0.7 V for MoS_2-WSe_2-EG and WSe_2-$MoSe_2$-EG, respectively) and then decreases to a minimum before undergoing a "hard" turn-on with exponential

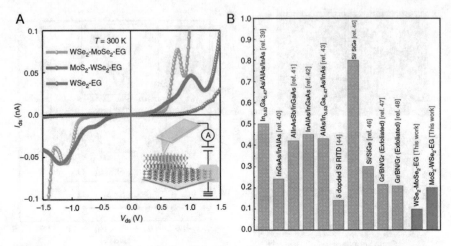

Fig. 7.6 Resonant tunneling and negative differential resistance in atomically thin layers. (**a**) Experimental *I–V* traces for different combination of dichalcogenide-graphene interfaces demonstrating NDR. The inset shows schematic of the experimental setup for the *I–V* measurement in this layered system. (**b**) Comparison of FWHM voltage of the NDR from this chapter with other reported results in room temperature [27, 38–47]

current increase (Fig. 7.6). The peak to valley current ratio (PVCR) is 1.9 for MoS_2-WSe_2-EG and 2.2 for WSe_2-$MoSe_2$-EG which is comparable to traditional resonant tunneling diodes [2–6].

7.3.5 NDR in vdW Heterostructures

Resonant tunneling between two spatially separated quantum states can be used to realize negative differential conductance. Negative differential conductance holds the key for novel nanoelectronic design options utilizing bistability and positive feedback. Novel memories, multivalued logic and inductor-free compact oscillators, and other electronic applications can benefit from a low-power, low-voltage negative differential conductance device. The resonant tunneling diode (RTD) has been a subject of intense study and design optimization, in silicon germanium and III–V heterostructure material systems for many years now. While theoretically capable of operating in extremely narrow voltage windows, the negative differential conductance of a RTD, particularly at room temperature, is limited by scattering mechanisms, related to interfacial imperfections, which are unavoidable even when utilizing high-vacuum advanced epitaxial growth technique. The interface-related scattering reduces the sensitivity of resonant tunneling to an external bias, thereby increasing the voltage window over which negative differential conductance regime is observed. van der Waals epitaxy of 2D materials can mitigate these issues and provide a material platform for device engineers to obtain energetically sharp NDR features at room temperature leading to novel low-power quantum tunneling devices.

7.4 Conclusions

In this chapter, we demonstrate the direct synthesis of unique multi-junction hetero-structures based on epitaxial graphene on SiC, MoS_2, $MoSe_2$, and WSe_2 that yields pristine interlayer gaps and leads to the first demonstration of resonant tunneling in an atomically thin synthetic stack with the spectrally narrowest room-temperature NDR characteristics. Importantly, this work indicates that NDR at room temperature only occurs in TMDC-based heterostructures with truly pristine interfaces, which has been recently corroborated with manually stacked heterostructures where NDR is only evident at liquid nitrogen temperatures [17, 48, 49]. This is due to reso-nant tunneling being highly sensitive to interfacial perturbations such as defects or "residue" from the transfer process, emphasizing the importance of direct synthesis of multi-junction TMDC heterostructures for vertical quantum electronics applica-tions. Interestingly, the room-temperature full width at half maximum (FWHM) of the NDR in this work is more spectrally narrow than their "3D" semiconductor counterparts (silicon, germanium, III–V) and manually stacked graphene-boron nitride-graphene (Gr-hBN-Gr) heterostructures (Fig. 7.6b) [38–47]. This suggests that the interface of the directly grown vdW heterostructures is superior to that of many previously RTD structures.

References

1. Novoselov, K.S., Mishchenko, A., Carvalho, A., Castro Neto, A.H.: 2D materials and van der Waals heterostructures. Science. **353**, 80 (2016)
2. Esaki, L.: New phenomenon in narrow Germanium p-n junctions. Phys. Rev. **109**, 603–604 (1958)
3. Esaki, L., Tsu, R.: Superlattice and negative differential conductivity in semiconductors. IBM J. Res. Dev. **14**, 61–65 (1970)
4. Mitin, V.V., Kochelap, V., Stroscio, M.A.: Quantum heterostructures: microelectronics and optoelectronics. Cambridge University Press, Cambridge (1999)
5. Chan, H.L., Mohan, S., Mazumder, P., Haddad, G.I.: Compact multiple-valued multiplexers using negative differential resistance devices. IEEE J. Solid-State Circuits. **31**, 1151–1156 (1996)
6. Bayram, C., Vashaei, Z., Razeghi, M.: AlN/GaN double-barrier resonant tunneling diodes grown by metal-organic chemical vapor deposition. Appl. Phys. Lett. **96**, 042103 (2010)
7. Novoselov, K.S., et al.: Electric field effect in atomically thin carbon films. Science. **306**, 666–669 (2004)
8. Novoselov, K.S., et al.: Two-dimensional atomic crystals. Proc. Natl. Acad. Sci. USA. **102**, 10451–10453 (2005)
9. Geim, A.K., Grigorieva, I.V.: Van der Waals heterostructures. Nature. **499**, 419–425 (2013)
10. Zhan, Y., Liu, Z., Najmaei, S., Ajayan, P.M., Lou, J.: Large-area vapor-phase growth and char-acterization of MoS_2 atomic layers on a SiO_2 substrate. Small. **8**, 966–971 (2012)
11. Lee, Y.-H., et al.: Synthesis of large-area MoS_2 atomic layers with chemical vapor deposition. Adv. Mater. **24**, 2320–2325 (2012)
12. Gutiérrez, H.R., et al.: Extraordinary room-temperature photoluminescence in triangular WS_2 monolayers. Nano Lett. **13**, 3447–3454 (2013)

13. Liu, K.-K., et al.: Growth of large-area and highly crystalline MoS_2 thin layers on insulating substrates. Nano Lett. **12**, 1538–1544 (2012)

14. Zhang, Y., et al.: Direct observation of the transition from indirect to direct bandgap in atomically thin epitaxial $MoSe_2$. Nat. Nanotechnol. **9**, 111–115 (2014)

15. Ugeda, M.M., et al.: Giant bandgap renormalization and excitonic effects in a monolayer transition metal dichalcogenide semiconductor. Nat. Mater. **13**, 1091–1095 (2014)

16. Terrones, H., López-Urías, F., Terrones, M.: Novel hetero-layered materials with tunable direct band gaps by sandwiching different metal disulfides and diselenides. Sci. Rep. **3**, 1549 (2013)

17. Fang, H., et al.: Strong interlayer coupling in van der Waals heterostructures built from single-layer chalcogenides. Proc. Natl. Acad. Sci. USA. **111**, 6198–6202 (2014)

18. Haigh, S.J., et al.: Cross-sectional imaging of individual layers and buried interfaces of graphene-based heterostructures and superlattices. Nat. Mater. **11**, 764–767 (2012)

19. Rivera, P., et al.: Observation of long-lived interlayer excitons in monolayer $MoSe_2$-WSe_2 heterostructures. Nat. Commun. **6**, 6242 (2015)

20. Chiu, M.-H., et al.: Spectroscopic signatures for interlayer coupling in MoS_2-WSe_2 van der Waals stacking. ACS Nano. **8**, 9649–9656 (2014)

21. Robinson, J.A., et al.: Epitaxial graphene transistors: enhancing performance via hydrogen intercalation. Nano Lett. **11**, 3875–3880 (2011)

22. Huang, J.-K., et al.: Large-area synthesis of highly crystalline WSe_2 monolayers and device applications. ACS Nano. **8**, 923–930 (2014)

23. Eichfeld, S.M., et al.: Highly scalable, atomically thin WSe_2 grown via metal-organic chemical vapor deposition. ACS Nano. **9**, 2080–2087 (2015)

24. Su, S.-H., et al.: Band gap-tunable molybdenum sulfide selenide monolayer alloy. Small. **10**, 2589–2594 (2014)

25. Ghosh, R.K., Lin, Y.-C., Robinson, J.A., Datta, S.: Heterojunction resonant tunneling diode at the atomic limit. 2015 International Conference on Simulation of Semiconductor Processes and Devices (SISPAD), 266–269. IEEE (2015).

26. Lin, Y.-C., et al.: Direct synthesis of van der Waals solids. ACS Nano. **8**, 3715–3723 (2014)

27. Lin, Y.-C., et al.: Atomically thin resonant tunnel diodes built from synthetic van der Waals heterostructures. Nat. Commun. **6**, 7311 (2015)

28. Lee, G.-H., et al.: Electron tunneling through atomically flat and ultrathin hexagonal boron nitride. Appl. Phys. Lett. **99**, 243114 (2011)

29. Gong, Y., et al.: Vertical and in-plane heterostructures from WS_2/MoS_2 monolayers. Nat. Mater. **13**, 1135–1142 (2014)

30. Wang, Q.H., Kalantar-Zadeh, K., Kis, A., Coleman, J.N., Strano, M.S.: Electronics and opto-electronics of two-dimensional transition metal dichalcogenides. Nat. Nanotechnol. **7**, 699–712 (2012)

31. Chiu, M.-H., et al.: Determination of band alignment in transition metal dichalcogenides heterojunctions (2014)

32. Yang, W., et al.: Epitaxial growth of single-domain graphene on hexagonal boron nitride. Nat. Mater. **12**, 792–797 (2013)

33. Parkinson, B.A., Ohuchi, F.S., Ueno, K., Koma, A.: Periodic lattice distortions as a result of lattice mismatch in epitaxial films of two-dimensional materials. Appl. Phys. Lett. **58**, 472–474 (1991)

34. Klein, A., Tiefenbacher, S., Eyert, V., Pettenkofer, C., Jaegermann, W.: Electronic band structure of single-crystal and single-layer WS_2: influence of interlayer van der Waals interactions. Phys. Rev. B. **64**, 205416 (2001)

35. Zhang, C., Johnson, A., Hsu, C.-L., Li, L.-J., Shih, C.-K.: Direct imaging of band profile in single layer MoS_2 on graphite: quasiparticle energy gap, metallic edge states, and edge band bending. Nano Lett. **14**, 2443–2447 (2014)

36. Rawlett, A.M., et al.: Electrical measurements of a dithiolated electronic molecule via conducting atomic force microscopy. Appl. Phys. Lett. **81**, 3043 (2002)

37. Georgiou, T., et al.: Vertical field-effect transistor based on graphene-WS_2 heterostructures for flexible and transparent electronics. Nat. Nanotechnol. **8**, 100–103 (2013)
38. Smet, J.H., Broekaert, T.P.E., Fonstad, C.G.: Peak-to-valley current ratios as high as 50:1 at room temperature in pseudomorphic $In_{0.53}Ga_{0.47}As/AlAs/InAs$ resonant tunneling diodes. J. Appl. Phys. **71**, 2475 (1992)
39. Day, D.J., Yang, R.Q., Lu, J., Xu, J.M.: Experimental demonstration of resonant interband tunnel diode with room temperature peak-to-valley current ratio over 100. J. Appl. Phys. **73**, 1542–1544 (1993)
40. Su, Y.-K., et al.: Well width dependence for novel AlInAsSb/InGaAs double-barrier resonant tunneling diode. Solid State Electron. **46**, 1109–1111 (2002)
41. Tsai, H.H., Su, Y.K., Lin, H.H., Wang, R.L., Lee, T.L.: P-N double quantum well resonant interband tunneling diode with peak-to-valley current ratio of 144 at room temperature. IEEE Electron Device Lett. **15**, 357–359 (1994)
42. Rommel, S.L., et al.: Epitaxially grown Si resonant interband tunnel diodes exhibiting high current densities. IEEE Electron Device Lett. **20**, 329–331 (1999)
43. See, P., et al.: High performance $Si/Si_{1-x}/Ge_x$ resonant tunneling diodes. IEEE Electron Device Lett. **22**, 182–184 (2001)
44. Jin, N., et al.: Diffusion barrier cladding in Si/SiGe resonant interband tunneling diodes and their patterned growth on PMOS source/drain regions. IEEE Trans. Electron Devices. **50**, 1876–1884 (2003)
45. Britnell, L., et al.: Resonant tunnelling and negative differential conductance in graphene transistors. Nat. Commun. **4**, 1794 (2013)
46. Evers, N., et al.: Thin film pseudomorphic $AlAs/In_{0.53}Ga_{0.47}As/InAs$ resonant tunneling diodes integrated onto Si substrates. IEEE Electron Device Lett. **17**, 443–445 (1996)
47. Mishchenko, A., et al.: Twist-controlled resonant tunnelling in graphene/boron nitride/graphene heterostructures. Nat. Nanotechnol. **9**, 808–813 (2014)
48. Roy, T., et al.: Dual-gated MoS_2/WSe_2 van der Waals tunnel diodes and transistors. ACS Nano. **9**, 2071–2079 (2015)
49. Lee, C.-H., et al.: Atomically thin p–n junctions with van der Waals heterointerfaces. Nat. Nanotechnol. **9**, 676–681 (2014)

Chapter 8
Summary

The work in this thesis was trying to probe the viability of producing high-quality and large-area 2D semiconductors, transition metal dichalcogenides in particular, by thin-film deposition. Many aspects, including the surface termination and crystal facets of substrates used for growth and the thin-film deposition conditions, are taken into account in order to achieve high-quality results. This integration engineering is carried out on two types of interfaces: 2D materials on 3D substrates (quasi-vdW epitaxy) and 2D materials on vdW substrates (regular vdW epitaxy).

From a perspective of material growth, the substrate properties play an important role for scalable and stoichiometric 2D layers. This aspect is in particular critical when the harsh environments and multiple chemical precursors are involved in MOCVD process and is discussed in Chap. 3. Using vdW solids as substrates for material growth can prevent 2D alloys from forming. This could be contributed to the lack of dangling bonds and simpler dielectric environment for molecules during the synthesis, compared to that of 3D substrates. In order to achieve the epitaxial quality, the structural coincidence between the 2D layers and the substrates chosen for growth is required. And this is well discussed in Chaps. 3, 4, and 7. A variety of characterization methods, including Raman and photoluminescence spectrum, electron microscopes, scanning probe microscope techniques, and also electrical measurements, have been employed in order to understand the results.

This thesis established a milestone for direct synthesis of 2D layers for heterogeneous integration for new functionality and nanoelectronics. In particular, we found the properties of the supporting substrates can impact the electronic transports of synthetic 2D layers. For example, Chaps. 5 and 6 show that the doping type and carrier concentration of EG can significantly modulate the electrical barrier between

© Springer Nature Switzerland AG 2018
Y. -C. Lin, *Properties of Synthetic Two-Dimensional Materials and Heterostructures*, Springer Theses,
https://doi.org/10.1007/978-3-030-00332-6_8

WSe$_2$ and EG. Similarly, back to the second section of Chap. 3, we show surface morphology and termination of sapphire affect the transport of FET devices of epitaxial WSe$_2$ significantly. Furthermore, we also prove that sophisticated heterostructures that go beyond only two stacked 2D layers can make possible through material growth; they can provide interesting properties that are not accessible through the mechanical transfer approach, like those we have discussed in Chap. 7.

Vita

Education

- Ph.D., Materials Science and Engineering, *Pennsylvania State University, PA, USA* (2017)
- M.S., Physics, *National Taiwan University, Taiwan* (2011)
- B.S., Physics, *National Cheng Kung University, Taiwan* (2009)

Research Experiences

Pennsylvania State University, University Park, PA (January 2013–August 2017)
Graduate Research Assistant

- Performed epitaxial growth of two-dimensional (2D) materials by chemical vapor deposition
- Studied properties of 2D and relevant nanomaterials via optical and electrical characterizations
- Fabricated and measured transistors and diodes of 2D materials

National Taiwan University, Taiwan (June 2009–June 2011)
Graduate Research Assistant

- Synthesized graphene and graphene oxide and studied their electrical properties
- Performed device fabrication
- Investigated the magneto-transport on the 2D electron system in GaAs/AlGaAs

© Springer Nature Switzerland AG 2018
Y. -C. Lin, *Properties of Synthetic Two-Dimensional Materials and Heterostructures*, Springer Theses,
https://doi.org/10.1007/978-3-030-00332-6

National Cheng Kung University, Taiwan (June 2007–June 2009)
<u>Undergraduate Research Assistant</u>

- Investigated the impact of the geometry of defective AFM tips on imaging $BiFeO_3$

Teaching and Mentorship

Pennsylvania State University, University Park, PA (Fall 2015 and 2016)
   "Metallurgical Laboratory" with the instructor, Amy C. Robinson

- Led two sessions weekly for a class of 20 undergraduates majored in Materials Sciences
- Prepared lab materials and taught lectures to students prior to the lab

   National Taiwan University, Taiwan (Fall 2009–Spring 2011)
   "Quantum Mechanics" at graduate level, with Professor Jeng-Wei Chen
   "Electromagnetism" at undergraduate level, with Professor Jeng-Wei Chen

- Engaged students in the office hours; graded exams and homework

   Student Researcher Mentees

- Jennifer G. DiStefano (Penn State undergraduate, January 2013–May 2016)
- Kursti S. DeLello (Univ. of Central Florida undergraduate, REU, May 2016– August 2016)
- Jing-Kai Huang (National Chiao-Tung Univ. master degree, May 2010– December 2011)

Award

- Graduate Student Award Silver Medalist, Materials Research Society, *USA* (2016)

Selected Publications

1. Y.-C. Lin, R. Addou, Q. Wang, C. Smyth, S. M. Eichfeld, Ganesh R. Bhimanapati, M. J. Kim, R. M. Wallace, J. A. Robinson, "Novel defect-mediated transport in van der Waals heterostructures" (Submitted 2019)
2. Y.-C. Lin,[†] B. Jariwala,[†] B. M. Bersch, K. Xu, Y. Nie, B. Wang, S. M. Eichfeld, X. Zhang, T. Choudhury, Y. Pan, R. Addou, C. M. Smyth, J. Li, K. Zhang, A. Haque, S. Fölsch, R. M. Feentra, R. M. Wallace, K. Cho, S. K. Fullerton,

J. M. Redwing, J. A. Robinson, "Large-area, Electronic-grade Two-dimensional Semiconductors", *ACS Nano* (2018) († Equal contribution)

3. B. M. Bersch,[†] S. M. Eichfeld,[†]Y.-C. Lin,[†] K. Zhang, G. R. Bhimanapati, A. F. Piasecki, M. Labella III, J. A. Robinson, "Selective area growth and controlled substrate coupling of transition metal dichalcogenides" *2D Materials* (2017), 4, 025083 († Equal contribution)

4. Y.-C. Lin, K. DeLello, H.-T. Zhang, K. Zhang, J. A. Robinson, "Photoluminescence of monolayer transition metal dichalcogenides integrated with VO_2" *Journal of Physics: Condensed Matter* (2016), 28, 50

5. Y.-C. Lin, J. Li, S. C. de la Barrera, S. M. Eichfeld, Y. Nie, R. Addou, P. C. Mende, R. M. Wallace, K. Cho, R. M. Feenstra, J. A. Robinson, "Tuning electronic transport in epitaxial graphene-based van der Waals heterostructures" *Nanoscale* (2016), 8, 8947

6. Y.-C. Lin, R. K. Ghosh, R. Addou, N. Lu, S. M. Eichfeld, H. Zhu, M.-Y. Li, X. Peng, M. J. Kim, L.-J. Li, R. M. Wallace, S. Datta, J. A. Robinson, "Atomically thin resonant tunneling diodes built from synthetic van der Waals heterostructures" *Nature Communications* (2015), 6, 7311

7. Y.-C. Lin, C.-Y. S. Chang, R. K. Ghosh, J. Li, H. Zhu, R. Addou, B. Diaconescu, T. Ohta, X. Peng, M. J. Kim, J. T. Robinson, R. M. Wallace, T. S. Mayer, S. Datta, L.-J. Li, J. A. Robinson, "Atomically thin heterostructures based on single-layer tungsten diselenide and graphene" *Nano Letters* (2014), 14, 6936–6941

8. Y.-C. Lin, N. Lu, N. Perea-Lopez, J. Li, Z. Lin, C. H. Lee, C. Sun, L. Calderin, P. N. Browning, M. S. Bresnehan, M. J. Kim, T. S. Mayer, M. Terrones, J. A. Robinson, "Direct synthesis of van der Waals Solids" *ACS Nano* (2014), 8, 3715–3723

Appendices

Appendix A

Parameters of MOCVD Process for Epitaxial WSe$_2$

A controlled layer-by-layer growth of WSe$_2$, ranging from monolayer to three layers, was achieved by following the growth profile illustrated in Fig. A.1. Growths were done at 700 Torr using H$_2$ as a carrier gas, where W(CO)$_6$ and H$_2$Se precursors are introduced separately into the cold wall vertical reactor chamber and their respective flow rates controlled via mass flow controllers (MFCs). The optimized condition for the growth was modified from our previously reported work [1]. In order to achieve uniform deposition with complete coalescence over the entire

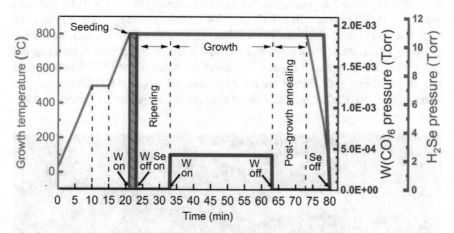

Fig. A.1 Optimized growth profile for MOCVD process of monolayer epitaxial WSe$_2$ grown on sapphire [2]

© Springer Nature Switzerland AG 2018
Y. -C. Lin, *Properties of Synthetic Two-Dimensional Materials and Heterostructures*, Springer Theses,
https://doi.org/10.1007/978-3-030-00332-6

substrate area, we introduce a step that combines seeding and annealing at 800 °C where the substrate is exposed to partial pressures of ~2 × 10⁻³ Torr W-precursor and 11 Torr H_2Se for 30 s to 2 min, named "pregrowth seeding step." Subsequently the WSe_2 is annealed in H_2Se to allow the nucleated domain to grow further under Se-rich environment. Also, such pre-annealing step at higher temperature acts as a surface treatment on c-sapphire surface and promotes Se passivation of the sapphire, which further acts as bridge for epitaxy at 2D/3D interface (detail can be found in main text). The number of layers was controlled by varying the seeding time, 30 s (1L) to 2 min (3L), while keeping growth time 30 min unchanged. All growths were done at 800 °C and 730 Torr total pressure with constant W:Se flux by adjusting the $W(CO)_6/H_2Se$ partial pressure individually for the following layer number: 4.32 × 10⁻⁴ Torr/10.8 Torr (1L), 6.24 × 10⁻⁴ Torr/15.6 Torr (2L), and 7.68 × 10⁻⁴ Torr /19.2 Torr (3L). Growth profile in Fig. A.1 is representative of deposition conditions for monolayer WSe_2 film.

Theoretical Modeling

The density functional theory (DFT) calculations were performed with the Vienna ab initio simulation package (VASP) [3]. The valence electronic states are expanded in a set of periodic plain waves, and the ion-electron interaction is implemented through the projector-augmented wave (PAW) approach [4]. The Perdew-Burke-Ernzerhof (PBE) GGA exchange-correlation functional is applied in the simulation [5]. The wave functions are expanded in plane waves with a kinetic energy cutoff of 400 eV. The convergence criteria for the electronic and ionic relaxation are 1.0 × 10⁻⁵ eV and 1.0 × 10⁻⁴ eV, respectively. Integration over the first Brillouin zone is performed with a Γ-centered 3 × 3 × 1 k-point mesh. A supercell consisting of 2 × 2 α-Al_2O_3 unit cells and 3 × 3 1L WSe_2 unit cells is built with a 4% strain on WSe_2. A vacuum layer of 20 Å is added to the c-direction. To avoid long-range interactions between supercells, a supercell consists of the surface and interface under study on both sides along the c-direction. The proposed WSe_2-sapphire interfaces based on the EDX data (Fig. A.2) are shown in Fig. A.3. Except for the Al-terminated sapphire/WSe_2 interface, gap states exist within the bandgap of WSe_2 after contact. Comparing the density of states

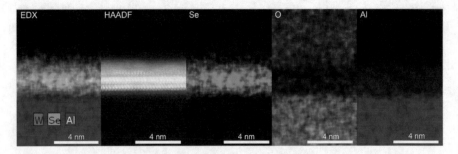

Fig. A.2 EDX and HAADF image captured from the same section during STEM measurement at 200 kV identify the distribution of W, Se, Al, and O [2]

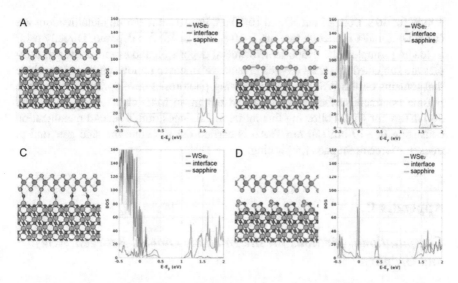

Fig. A.3 DFT modeling for 2D/3D interface. The proposed interface structures and their corresponding density of state (DOS) under different terminations: (**a**) Al-terminated (**b**) Al-Se-terminated (**c**) O-terminated (**d**) Al-O-Se-terminated. The 0 eV is the valence band edge for an intrinsic semiconductor [2]

of the gap states, the four interfaces are ordered as Al-O > Al-Se > Al-O-Se > Al. The calculations indicate that the interaction (bonding) energy between WSe_2 and the Se-terminated sapphire surfaces (4.23 eV for Al-Se connection in Fig. A.3b and 2.6 eV for Al-O-Se connection in the d) lies between that of WSe_2/Al-terminated (0.04 eV) and WSe_2/Al-O-terminated surfaces (5.4 eV). This relatively high interface bonding energy between WSe_2 and Al-Se connection also manifests itself mechanically, as we find that fully coalesced epitaxial WSe_2 layers are more difficult to mechanically transfer from the substrate than non-epitaxial WSe_2.

Device Fabrication

Field-effect transistors were fabricated via standard photolithography to define WSe_2 channel dimensions, source/drain (S/D) contact electrodes, and side-gate electrodes (Fig. 3.14). The 4.1 × 2.5 mm die layout employed in this work consists of an array of FETs with channel width 24 μm and channel length ranging from 10 μm to 0.75 μm. With these die dimensions in mind, a 3 row × 2 column die layout is used to cover a majority of the 10 × 10 mm sample surface. In our work, the gate electrode is not directly deposited on top of the electrolyte-WSe_2 FETs, and instead, we utilize a side-gate geometry that establishes a lateral electric field in the PEO:$CsClO_4$ (PEO: poly(ethylene-oxide)) and drives the ions into place on the WSe_2 channel surface. All photolithography was carried out in a GCA 8500 i-line stepper. WSe_2 channels were isolated and etched via reactive ion etching in a Plasma Therm PT-720 plasma etch tool using an $SF_6/O_2/Ar$ gas chemistry at 10 mTorr and

100 W for 30 s. Both 25 nm Ni and 10/10 nm Pd/Au source/drain metallizations are carried out under moderate vacuum (~10^{-6} Torr) at 1.0 Å/s dep rate. Directly prior to loading samples into evaporator for metal deposition and eventual lift-off, samples are subjected to a brief oxygen plasma treatment to remove photoresist residue that remains on the WSe$_2$ surface following photoresist development. This gentle plasma treatment/surface prep is carried out in an M4L etch tool at 50 W and 500 mTorr for 45 s. Following this initial metal deposition, a second metallization consisting of ~10 nm/150 nm Ti/Au is carried out to define the side gate and to thicken source/drain pads for probing.

Appendix B

Computational Methods for the Intrinsic Dipoles Between WSe$_2$ and Graphene

The density functional theory (DFT) calculation is performed by Vienna ab initio simulation package (VASP) [3] with the projector-augmented wave (PAW) method [4]. The local density approximation (LDA) [6] is used to describe the exchange-correlation functional with the partial core correction included. Spin polarization and spin-orbit coupling are applied. The stable phase of the monolayer WSe$_2$ is trigonal prism structure [7]. The optimized planar lattice constant of WSe$_2$ is 3.25 Å, and the optimized planar lattice constant for monolayer graphene is 2.45 Å. In order to fit the lattice constant, a supercell with 3 × 3 WSe$_2$ unit cells and 4 × 4 graphene unit cell is used, and a compressive strain of 0.4% is applied to graphene, as the electronic behaviors of TMDC are very much susceptible to lattice strain. The supercell is shown in Fig. B.1a. The wave functions are expanded in plane waves with a kinetic energy cutoff of 500 eV, and the convergence criteria for the electronic relaxation are 10^{-5} eV. Integration over the Brillouin zone is performed with a gamma-centered 6 × 6 × 1 Monkhorst-Pack k-point mesh for ionic and electronic optimization. A vacuum region of about 15 Å normal to the surface is added to minimize the interaction between adjacent slabs (Fig. B.1a). Dipole correction on the stacking direction is used in systems to reveal the dipole within the two layers caused by the Fermi-level alignment. The local density approximation (LDA) is found to be suitable for studying the metal-TMDC contact [8]. The generalized gradient approximation (GGA) [5] with the DFT-D2 method for van der Waals (vdW) corrections [9] is also used to cross-check the structural accuracy. We find that GGA results with vdW corrections are in overall agreement with LDA results. Both the LDA method and the GGA + vdW method result in a similar structure with a distance of ~3.5 Å between graphene and TMDC, indicating a secondary bond interaction. The energy difference between the vacuum regions on the both sides of the contact system is the dipole induced by the contact. The vacuum energy level above WSe$_2$ is 0.17 eV higher than that above graphene, indicating a dipole from graphene toward the WSe$_2$ (Fig. B.1b).

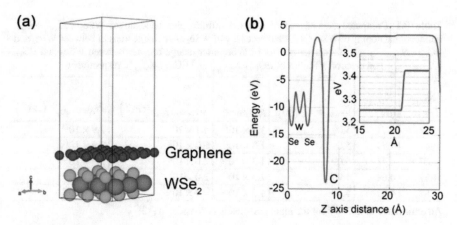

Fig. B.1 (**a**) Plane averaged local electric potential energy of electrons along the stacking direction. (**b**) After dipole correction, a difference on vacuum energy above both sides of 0.17 eV is observed (zoomed inset) [10]

Computation of WSe$_2$ Doping Density and Charge Densities and Dependence on Parameters

For the computation of charge transfer and band alignment, we take the doping densities of EG$_{PH}$ and EG$_{FH}$ from our experimental values, as discussed in the main text. Parameters in the computation are the electron affinities for monolayer and bilayer graphene, with nominal values of 4.57 eV and 4.71 eV, respectively, as known from prior experiments [11]. We take the sum of the electron affinity plus bandgap of the WSe$_2$, $X_{WSe_2} + E_g$, to be an unknown in the computation, since a value for this sum is not accurately known from prior work (only the sum is considered here since the electron occupation in the conduction band of the WSe$_2$ is negligible). A second unknown is the unintentional doping density of WSe$_2$. Then, using the two measured work function differences for WSe$_2$ on both EG$_{PH}$ and EG$_{FH}$ compared to the bare EG$_{PH}$ and EG$_{FH}$, we can determine values for the two unknown parameters. The carrier densities for the WSe$_2$ on both EG$_{PH}$ and EG$_{FH}$ after charge transfer are then a byproduct of the computation. In all cases, the carrier densities of WSe$_2$ in WSe$_2$-EG$_{PH}$ are very much greater than those of WSe$_2$ in WSe$_2$-EG$_{FH}$, consistent with the observed differences in the CAFM I–V results.

We note that the doping density values in Table B.1 are all the same, reflecting a tight constraint on this value. This constraint arises from charge transfer between the WSe$_2$ and the EG$_{PH}$. As pictured in Fig. B.2a, b, since the Fermi energies of the EG$_{PH}$ and WSe$_2$ are relatively far apart prior to charge transfer, and hence the Fermi energy of the WSe$_2$ ends up well within its bandgap after the transfer, then the p-type doping density in the WSe$_2$ is directly determined by the doping density of the EG together with the difference between the electron affinity of the EG$_{PH}$ and the $X_{WSe_2} + E_g$ value of the WSe$_2$. The resulting carrier densities for the WSe$_2$ on EG$_{PH}$ are negligible, again since the resulting WSe$_2$ Fermi energy is well within the gap. On the other hand, for the WSe$_2$ on EG$_{FH}$, their Fermi energies are relatively close prior to charge transfer, as pictured in Fig. B.2c, d. The resulting Fermi energy for

Table B.1 Computed dependence of electron affinity plus bandgap of WSe_2 ($X_{WSe_2} + E_g$), unintentional doping of WSe_2 (N_A), carrier density of WSe_2 after charge transfer between WSe_2 and EG_{PH} ($N_{C,WSe_2-EG_{PH}}$), and carrier density of WSe_2 after charge transfer between WSe_2 and EG_{FH} ($N_{C,WSe_2-EG_{FH}}$) on electron affinities of EG_{PH} ($X_{EG_{PH}}$) and EG_{FH} ($X_{EG_{FH}}$), respectively

$X_{EG_{PH}}$ (eV)	$X_{EG_{FH}}$ (eV)	$X_{WSe_2} + E_g$ (eV)	N_A (cm^{-2})	$N_{C,WSe_2-EG_{PH}}$ (cm^{-2})	$N_{C,WSe_2-EG_{FH}}$ (cm^{-2})
4.57	4.71	5.09	1.3×10^{12}	4.1×10^5	2.9×10^{12}
4.47	4.71	5.09	1.3×10^{12}	0.9×10^4	2.9×10^{12}
4.67	4.71	5.09	1.3×10^{12}	2.0×10^7	2.9×10^{12}
4.57	4.61	4.99	1.3×10^{12}	2.0×10^7	2.9×10^{12}
4.57	4.81	5.19	1.3×10^{12}	0.9×10^4	2.9×10^{12}

An error range of ±0.1 eV for the input parameters is considered [12]

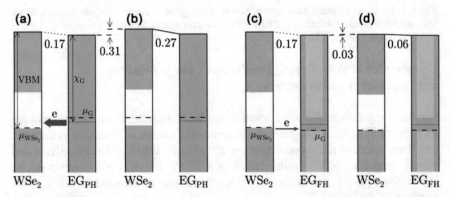

Fig. B.2 Band alignment of WSe_2 and EG_{PH} (**a**) before charge transfer (including computed intrinsic dipole 0.17 eV) and (**b**) after charge transfer. Band alignment of WSe_2 and EG_{FH} (**c**) before charge transfer (including the intrinsic dipole) and (**d**) after charge transfer. Monolayer and bilayer graphene models are employed for EG_{PH} and EG_{FH}, respectively, based on LEEM observations. Green shades in (**c**) and (**d**) represent conduction/valence subbands of bilayer graphene. The numerical values show various vacuum level differences, in units of eV [12]

the WSe_2 on EG_{FH} ends up near or within the valence band even after the charge transfer, with concomitant large carrier density, and the value of the WSe_2 doping density is not so tightly constrained in this part of the problem.

We have also considered the effect on the computed carrier densities of variation in the EG_{PH} and EG_{FH} doping density values, as well as variation of the measured work function differences within their experimental error ranges. Doping densities of $(4 \pm 1) \times 10^{12}$ cm^{-2} for EG_{PH} and $(1.5 \pm 0.2) \times 10^{13}$ cm^{-2} for EG_{FH} are typical measured in our samples. Considering the variations of these doping densities, the carrier density of WSe_2 on EG_{FH} after charge transfer is computed to range from 2.5 to 3.0×10^{12} cm^{-2}, while the carrier density of WSe_2 on EG_{PH} after transfer is always less than 10^7 cm^{-2}, i.e., its Fermi is well within the bandgap. For the measured error ranges (±0.03 eV) on the work function differences, performing computations at the bounds of these values produces carrier densities in the WSe_2 on EG_{FH} compared to WSe_2 on EG_{PH} that continue to differ by more than a factor of 10^4, for all cases.

Appendix C

Theoretical Validation for NDR Transport in the Trilayer Structures

We perform non-equilibrium ballistic quantum transport calculations by combining density functional theory (DFT) with the non-equilibrium Green's function (NEGF) formalism that provide theoretical *I–V* curves to confirm the NDR transport mechanism in the heterostructure by comparing it against the simulated transport in the homo-structure (Fig. C.1). In the experimental setup, the voltage, V_{ds}, is applied between the *Pt-Ir* tip of the conducting AFM and the electrically grounded graphene electrode. The area of the *Pt-Ir* tip is approximately to 1000 nm², which in the simulation is modeled as a bulk electrode in the theoretical structure (Fig. C.1a). The calculation produces the bias and the transverse momentum-dependent transmission probability of the carriers tunneling through the heterostructure and is used to simulate the *I–V* characteristics using Landauer transport formulation [13]:

$$I\left(V_{ds}\right) = \frac{2q}{h} \int_{BZ} dk \int dE T\left(E,, k,, V_{ds}\right) \left[f\left(\frac{E - E_{f_1}}{k_B T}\right) - f\left(\frac{E - E_{f_2}}{k_B T}\right) \right], \quad (C.1)$$

where $E_{f_1} - E_{f_2} = qV_{ds}$ represents the Fermi window, BZ represents the Brillouin zone, and $T(E, k\|, V_{ds})$ is the total transmission over the energy channels within the Fermi window calculated self-consistently for each V_{ds}. Within the NEGF + DFT framework for transport, the Hamiltonian of the system is solved by calculating the electronic charge distribution via the self-consistent DFT loop of the full density matrix of the device whose diagonal element describes the charge density.

This procedure produces the bias-dependent transmission function, $T(E, V, k\|)$. We then extract the *I–V* characteristics in the ballistic transport regime which shows a pronounced NDR in both positive and negative bias regimes of the MoS_2-WSe_2-Gr heterostructure device (Fig. C.1b). Within the Fermi window of 0–0.4 eV, we can see that the carrier transmission is effectively negligible due to the absence of any transmission channel. Above 0.4 eV, the transmission becomes finite, and the current starts increasing with the applied bias, where the primary transmission resonance peaks (peak P1, P2, and P3 in Fig. C.1c) appear at approximately 1.0 V and then get suppressed with further increase in applied bias. It is this peak and valley in the transmission spectra arising from resonant tunneling phenomenon that leads to the observed NDR. When the bias is further increased, conventional tunneling occurs due to the high density of states (DOS) at higher energy levels, and the current increases exponentially thereafter. The transmission Eigen states at the energetic location of the three strong peaks for a bias of +1.0 V provide clue to the microscopic origin of the NDR in the MoS_2-WSe_2-Gr heterostructure. Inspection of the localized molecular orbitals of the Eigen states (Fig. C.1d) reveals that all three resonance peaks originate from a combination of the Pt electrode (*s*-orbital), WSe_2 (*p*-orbital of Se, W and *d*-orbital of W), and graphene layers (*p*-orbital).

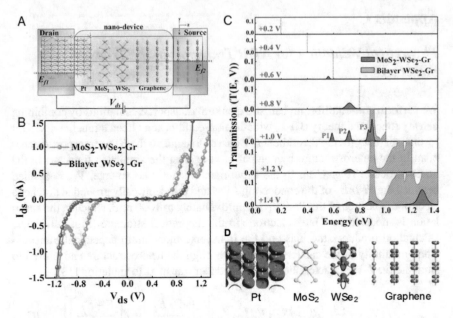

Fig. C.1 (**a**) Schematic of the vertical nano-device setup of both of MoS₂-WSe₂-Gr and bilayer WSe₂-Gr system used for quantum transport calculation. E_{f_1} and E_{f_2} indicate the corresponding Fermi levels of the left and right electrodes, respectively, for an applied positive bias V_{ds}. (**b**) Theoretical I–V curves of the vertical tunnel junctions for both the hetero- and homo-junction are simulated by the DFT and NEGF transport formalism that give resonant tunneling at specific energies and bias voltage, as shown in (**c**). The dotted line indicates the Fermi window for that applied bias voltage across the terminals. (**d**) Transmission Eigen states that contribute to the transmission in the peak P3 of the transmission at $V_{ds} = +1.0$ V in the MoS₂-WSe₂-Gr heterostructure [14]

In the case of MoS₂-WSe₂-Gr heterostructure, the MoS₂ in direct contact with the Pt electrode and the first graphene layer closest to the WSe₂ do not contribute to the strong transmission peaks but serve as tunnel barriers. Furthermore, the interatomic electronic interaction between the 2D layers makes MoS₂ n-type and WSe₂ p-type, which make the WSe₂ valence band states as the host for the confinement of the resonant states when the system is subjected to a bias. Along with the conservation of transverse momenta and the alignment of energy levels in the constituent layers of the system, the theoretical I–V traces are in good agreement with the measured results. On the other hand, bilayer WSe₂ does not offer any band offset in the energy band diagram, and its bandgap acts as a regular electronic barrier in the carrier tunneling. The calculated transmission in bilayer WSe₂-Gr clearly reflects this nature and shows no NDR in its I–V characteristics. This study hence provides strong theoretical insights that show resonant tunneling is the dominant transport mechanism in a heterostructure with significant amounts of band offset.

References

1. Eichfeld, S.M., et al.: Highly scalable, atomically thin WSe_2 grown via metal-organic chemical vapor deposition. ACS Nano. **9**, 2080–2087 (2015)
2. Lin, Y.C., et al.: Realizing large-scale, electronic-grade two-dimensional semiconductors. ACS Nano. **12**, 965–975 (2018)
3. Kresse, G., Furthmüller, J.: Efficient iterative schemes for ab initio total-energy calculations using a plane-wave basis set. Phys. Rev. B. **54**, 11169–11186 (1996)
4. Kresse, G., Joubert, D.: From ultrasoft pseudopotentials to the projector augmented-wave method. Phys. Rev. B. **59**, 1758–1775 (1999)
5. Perdew, J.P., Burke, K., Ernzerhof, M.: Generalized gradient approximation made simple. Phys. Rev. Lett. **77**, 3865 (1996)
6. Ceperley, D.M., Alder, B.J.: Ground state of the electron gas by a stochastic method. Phys. Rev. Lett. **45**, 566–569 (1980)
7. Gong, C., et al.: Band alignment of two-dimensional transition metal dichalcogenides: application in tunnel field effect transistors. Appl. Phys. Lett. **103**, 053513 (2013)
8. Gong, C., Colombo, L., Wallace, R.M., Cho, K.: The unusual mechanism of partial Fermi level pinning at metal-MoS_2 interfaces. Nano Lett. **14**, 1714–1720 (2014)
9. Makov, G., Payne, M.: Periodic boundary conditions in ab initio calculations. Phys. Rev. B. **51**, 4014–4022 (1995)
10. Lin, Y.-C., et al.: Tuning electronic transport in epitaxial graphene-based van der Waals heterostructures. Nanoscale. **8**, 8947–8954 (2016)
11. Yu, Y.-J., et al.: Tuning the graphene work function by electric field effect. Nano Lett. **9**, 3430–3434 (2009)
12. Lin, Y.-C., et al.: Wafer-scale MoS_2 thin layers prepared by MoO_3 sulfurization. Nanoscale. **4**, 6637 (2012)
13. Datta, S.: Quantum transport-atom to transistor. Cambridge University Press, Cambridge (2005)
14. Ghosh, R.K., Lin, Y.-C., Robinson, J.A., Datta, S.: Heterojunction resonant tunneling diode at the atomic limit. International Conference on Simulation of Semiconductor Processes and Devices (SISPAD), 266–269. IEEE (2015)

© Springer Nature Switzerland AG 2018
Y. -C. Lin, *Properties of Synthetic Two-Dimensional Materials and Heterostructures*, Springer Theses,
https://doi.org/10.1007/978-3-030-00332-6

Printed in the United States
By Bookmasters